THE POWER TO FLY
an engineer's life

THE POWER TO FLY
an engineer's life

Brian H. Rowe
Chairman Emeritus
General Electric Aircraft Engines

with Martin Ducheny

American Institute of Aeronautics and Astronautics, Inc.
1801 Alexander Bell Drive
Reston, Virginia 20191–4344
Publishers since 1930

American Institute of Aeronautics and Astronautics, Inc., Reston, Virginia

1 2 3 4 5

Library of Congress Cataloging-in-Publication Data

Rowe, Brian H., 1931-
  The power to fly: an engineer's life / Brian H. Rowe with Martin Ducheny.
     p. cm.
  Includes index.
  ISBN 1-56347-709-2 (alk. paper)
 1. Rowe, Brian H., 1931- 2. Aeronautical engineers--United States--Biography. 3. GE Aircraft Engines (Firm)--History. I. Ducheny, Martin. II. Title.

  TL540.R67A3 2004
  629.13'0092--dc22

2004011836

Book design by Chris McKenzie
Cover design by Gayle Machey

Copyright © 2005 by the American Institute of Aeronautics and Astronautics, Inc. All rights reserved. Printed in the United States of America. No part of this publication may be reproduced, distributed, or transmitted, in any form or by any means, or stored in a data base or retrieval system, without the prior written permission of the publisher.

Data and information appearing in this book are for informational purposes only. AIAA is not repsponsible for any injury or damage resulting from use or reliance, nor does AIAA warrant that use or reliance will be free from privately owned rights.

Foreword

One thing aerospace leaders share is a passion for flight. The dream to fly higher, faster, and farther has driven our finest engineering and science talents to achieve what many people thought was impossible.

Brian Rowe is one of those leaders. As a young boy in England, he had a front-row seat to some of aviation history's greatest events. And later, as a successful engineer and business leader, he witnessed the beginning of jet aviation and contributed to some of our industry's most important technology advances. His story is a fascinating personal account, as well as an interesting and compelling history of General Electric's jet engine business and the competitive business environment we operate in. Today's leaders would be wise to study and learn from his experiences and management philosophy, which Brian generously shares with us all.

Brian traces the growth of early military engines, such as GE's J47, that powered the world's first large, swept-wing jet—the Boeing B-47—and follows that to engines that powered other famous Boeing-built military planes, such as the F-4 Phantom and F/A-18 Hornet. The most important development in commercial aviation after the swept wing follows: high-bypass engines that made the Boeing's 747 and 777, and now our super-efficient 7E7 Dreamliner, possible.

One of the most interesting and, for me, personally relevant parts of Brian's story is about the GE90 engine, an engineering marvel just like the plane it was intended for—the 777. GE faced unprecedented technology and business challenges in developing this revolutionary engine. From my perspective, the company's experience reflects the challenges and opportunities facing everyone in our industry. We strive to bring more value to our customers by creating new and innovative products to meet their needs and expectations, often with limited development funds. And the global nature of our business offers us all the rewarding opportunity to work with diverse partners and suppliers from around the world to accomplish our goals, which is one reason our approach to designing the 777 was so successful.

Another reason for our success was our focus on delivering quality. The amazing amount of power demanded of modern jet turbine engines has led to a very complex machine that operates in an environment of incredible forces, pressure, and temperatures. Producing these new-generation engines requires unprecedented precision and quality. Brian was a catalyst in bringing continuous quality improvements to GE's Engine Division, where he experienced first-hand the challenge of adapting and remaining relevant in a changing and uncertain business environment. He knows the importance of maintaining a key technology base during a downturn, something GE was able to do through its Centers of Excellence.

Business success depends on great leadership, and Brian gives us the benefit of his experience and management philosophy, which is really quite simple: be passionate about what you do, value teamwork, learn from losing, focus on being a winner, and most important, appreciate and respect the people who support you at work and at home.

Foreword

I enjoyed this book very much. The history of GE's engine program is fascinating, and Brian's enthusiasm and intimate technical knowledge made me appreciate even more the talents of the men and women who give so much to make our world the best it can be with airplanes that connect us efficiently and safely. Brian's personal story gave me additional insight into a man I greatly respect, and someone who I believe is an example tomorrow's leaders can follow, and today's leaders can learn from.

Alan R. Mulally
Executive Vice President, The Boeing Company
President and Chief Executive Officer, Commercial Airplanes

TABLE OF CONTENTS

Preface xi
Photo Acknowledgments xiii

Chapter 1 Portrait of the Engineer as a Young Man 1
    A sign from above 1
    The ties that bind—setting the stage 2
    Head in the clouds 3
    On the playing fields 5
    Life as an apprentice 6
    Craftsmanship 9
    Designing ways 10
    The Goblin 12
    The Ghost and the Comet 12
    The Gyron 14
    The university and beyond 15
    Making rockets 16

Chapter 2 Coming to America—and to General Electric 19
    Our foreign adventure 19
    The Alien Office 20
    Lift fans 21
    The GE1 23
    Moving to Lynn 24
    The legacy of the lift fan 25
    Adopted by the East Coast establishment 26

Chapter 3 Earning My Stripes—Leading Engineers 27
    A French connection for the CF700 28
    Lynn Production Engineering 29
    Gerhard Neumann 32
    Summoned to Cincinnati 33
    Lynn's contribution to the growth of GE Aircraft Engines 34

Chapter 4 Reentering the Large Commercial Engine Market 39
    GE's large commercial engine business: a concise history 40
    The breakthrough technology of the TF39 42
    The commercial engine market of the mid-1960s 46
    Tri-jets: the Douglas DC-10 and the Lockheed L-1011 47
    Selling the CF6 48
    Selling the DC-10 in Europe 50
    Customer support as a competitive advantage 52
    Airbus Industrie 54

Engines for Boeing	56
The CF6-80 Series	57

Chapter 5 Love and Hate—Working with the Military 59

When your ideas are not your own	60
Up the bureaucracy	60
Myopic focus on cost	62
Playing both ends against the middle	63
From military to commercial	64
A changing attitude	66
The Great Engine War	66

Chapter 6 Swords into Plowshares—the CFM56 Family 69

Snecma goes a'courting	69
The birth of the CFM56	70
In search of a customer	71
New engines for old—the KC-135 tanker	72
Gambling on the Boeing 737	73
Airbus Industrie	74
Squeezing out every drop	74

Chapter 7 Other Engines: M&I and UDF Non-Flying Gas Turbines 77

M&I Department	77
Unducted Fan—the Ultra High Bypass Engine	79

Chapter 8 Leadership in a Large Technology Business 81

Picking people	84
Leading a team	85
Applying Neumann's axioms	86
Learning from Jack Welch	88
Inspired by Rene Ravaud	90
Direction, consistency, respect, and communication	91
The GE management system	92

Chapter 9 Manufacturing—The Hidden Heroes 97

Early lessons	97
Materials, processes, and people	98
Special processes	102
Staying number one	104

Chapter 10 Lighting the Torch of Quality 107

A culture committed to inspection	107
Compartmentalization	108
Benchmarking	108
Top down	109

Table of Contents

Bottom up	109
Consultants	110
Continuous Improvement vs Workout	110
Everybody's someone's supplier	111
Modules and commodities	112
Six Sigma Quality	113

Chapter 11   Overcoming a Failure of Integrity — 115
- The Dotan affair — 115
- A punishing lesson — 116

Chapter 12   Going Down to Go Up — 119
- Downsizing — 119
- The bottom falls out — 120
- A philosophical conflict — 121
- Making money — 123
- Stepping into the light — 123
- Lessons — 124

Chapter 13   The Lessons and Legacy of the GE90 — 127
- A technological challenge — 128
- The first competitions — 129
- British Airways — 130
- GE and Rolls-Royce — 132
- MTU slips away — 133
- Funding crisis — 134
- More technological challenges — 135
- Leading the development program — 142
- First flight — 143
- Going into service — 144
- The legacy — 145
- A view toward the future — 145

Chapter 14   Politics and the Engineer — 149
- Clement Atlee — 150
- Lobbying for the sale — 150
- International intrigues — 151
- Habibie and Indonesia — 153
- The inscrutable East — 156
- The King of Morocco — 158
- Bush the Elder — 158
- The U.K. and British Airways — 160
- United Technologies — 161
- Environmentalists — 161

Chapter 15   My Sporting Life—Some Side Thoughts   165
   Soccer   165
   Cricket   166
   Basketball   166
   Swimming   166
   Tennis   167
   Golf   168
   Lessons   168

Chapter 16   The Conquistadores   171
   Risking life and limb   171
   Selective membership   172
   Pride in leadership   173

Chapter 17   Values in the Turmoil of Leadership   175
   How to be lucky   176
   Family support   177
   Philosophy of life   178

Chapter 18   Life after GE   181
   Fifth Third Bank   182
   Aerostructures Hamble Ltd.   183
   Stewart & Stevenson   183
   Atlas Air   184
   B/E Aerospace   186
   Textron   187
   Canadian Marconi   188
   Cincinnati Bell   188
   Dynatech/Acterna   189
   Fairchild-Dornier   190
   Working with my son   191
   Being on boards   191

Epilogue: The Propulsion Hall of Fame   193

Index   197

Supporting Materials   207

Preface

When I began writing this book, I expected to put together no more than a remembrance of my career that I could give my children and grandchildren—something that might tell a few interesting stories and explain a bit of who I was and how I became the man I did. But it grew. As this work expanded, I realized how much being an engineer was an essential part of my life story, and I expanded my expectations. I hoped that this book might also encourage young people who were about to embark on their paths in the world to consider engineering as their own good start. When I looked over the whole story playing out, I recognized another significant facet. Over the years I learned important lessons about the grand world of business and how to be a leader in that world. I felt it would be a good thing to share those lessons with all who wished to learn from them.

I intended this to be my story, and so it is. To tell that story, I relate quite a bit of history, but I must warn the reader that this is in no way intended to be a complete history of GE's jet engine business during the years I knew it. Many engineering and business achievements are glossed over or not mentioned at all. As conscientiously as I tried to remember all of the important people and events that shaped the world I passed through, I am sure I forgot more than a few. In fact, there were probably many I was not even aware of at the time. In my narrative I describe conversations that happened years ago. I did not keep reporter's notes or make tape recordings of my conversations. What I present is my best recollection of the sense of what was said, even if I present them as quotations. So much for caveats.

Three things stood out for me as I looked over this story of my life. The first is how lucky I have been. I did start with some native ability, I did get a good education, and I did work hard, but that can be said of many people. The second is that I had a great family—my mother, father, and sister of course, but most especially my wife, Jill, and our kids, Linda, Penny, and David, who put up with me as I traveled long distances and worked long hours to turn dreams into reality. Yet only a few have experienced the success I have. I can only ask, "Why me?" No answer comes. The third humbling thought that struck me was how much my success depended on the good work of an impressive cadre of dedicated people. Without belittling what I have accomplished, I freely admit I did nothing alone. I am deeply indebted to so many who have worked with me or influenced me by their examples, and I hope I make that clear as the story unfolds.

The task of telling this story of one engineer's life was a protracted one. Now that it is completed, I am tempted to draw some succinct meaning from it to lay before you as a preamble. What can I say about this story that might encourage you to read on with enthusiasm? First, it is a tale of passion—the love of and commitment to a business that was energized by a drive to create an ever better tomorrow through engineering. And it is about people—the very best people—people who share this passion to deliver the best. Of course, it a story of products—the very best jet engines that design and manufacturing technology could conceive and produce coupled with the finest support that ensured satisfied customers. It is also a tale of resolute persistence in the endless struggle to overcome

With my grandson, Nick, at Aircraft Engines family day in 1988.

inertia and bureaucracy—refusing to rest on good enough. And finally it is a story of pride—personal pride in the tireless team, in the remarkable products, and in a job well done—pride that fueled my life and the lives of people around me. Passion, people, products, persistence, pride—quite a story. And, oh yes, we also made some money while we were having all of this fun. That is my take anyway. Maybe after you hear the whole story, you will find something else.

Brian H. Rowe
June 2004

PHOTO ACKNOWLEDGMENTS

Photos courtesy of General Electric on pages xi, 11, 14, 22, 27, 31, 35, 37, 43, 44, 45, 48, 51, 52, 67, 75, 78, 80, 91, 99, 103, 122, 129, 136, 138, 139, 140, 142, 159, 194.

Photos provided by the author on pages 7, 82, 89, 172, 173, 177, 181.

Photo courtesy of Dassault on page 29.

Photo courtesy of U.S. Army Aviation Museum Archive on page 23.

Photo courtesy of British Airways on page 144.

CHAPTER 1
Portrait of the Engineer as a Young Man

A sign from above

It all began in England. I remember a clear blue early autumn day in 1940 in Edgeware, a suburb of London about 10 miles northwest of the heart of the city. The Battle of Britain was raging in the skies, but it had little effect on a boy of 10. War seemed far away. Close at hand was the adventure of finding pieces of shrapnel from the flak guns to show off at school—or memorizing the silhouettes of friendly and enemy aircraft.

I knew them all, friend and foe—and spent time drawing pictures of glorious flaming dogfights rather than attending to lessons. Toys were hard to come by, so I would build balsa wood models of my favorite aircraft and share them with friends.

But this afternoon was to be special. I had been out riding bicycles with my Uncle Bob, who was more like an older brother to me. (Bob's luck was good. He ultimately survived being torpedoed twice in the North Sea as the war progressed.) As we looked up, we spotted a flight of Hawker Hurricanes heading home. One was trailing a ribbon of ominous smoke. We got off our bikes and watched.

The wounded Hurricane soon fell behind the formation, banked, and began losing altitude. As we watched, the pilot opened the canopy, jumped from the aircraft, and opened his parachute. The dying airplane continued on its course—straight at us.

In an instant, we went from being observers to participants. After a moment that seemed frozen in time, we ran with our bikes—for some unknown reason directly at the approaching plane. The Hurricane grew huge as it passed directly overhead. Just beyond us, it grazed a house, knocking the chimney off and rupturing the plane's oil tank. (I later learned that the sister of a friend of mine had been burned by the spraying oil.)

About a hundred yards past us, the Hurricane hit the ground in an empty lot and skidded on its belly to a halt. The impact had apparently shorted the machine gun firing control, and the plane's guns opened fire on a model house at the end of the street, blasting away in a sort of death spasm until it had expended all of its ammunition—and had totally destroyed the house.

Meanwhile, the pilot, a Polish national of the Polish auxiliary fighting side-by-side with the Royal Air Force, drifted to earth. Members of the home guard, men too old or unfit for the call that had put most of Britain's men in military service, armed themselves with pikes—there were not enough guns to arm them—and went in pursuit of this destroyer of the English homeland. Luckily, the pilot, who spoke no English and looked believably foreign, landed in the courtyard of a convent. The nuns used their persuasive skills to save the Polish flyer from the rough justice of the home guard with their blood up.

What is important about this incident is not that it sparked in me an essential obsession with flight. It did not; I had been hooked long before this, although it did help confirm that burning fire. The meaning that I carry from that Hurricane falling into my life is one of luck. As I look back from the vantage of a long, prosperous career in the aerospace business, it seems like a wonderful dream when I consider that it began with a boyhood fascination born of watching biplanes at air shows and building airplane models from scratch. I cannot help but think that it was luck as much as anything else that brought me to this point.

That errant fighter plane could have come to earth a few hundred feet earlier in its death glide, and my story might have ended there. In fact, the whole scene is a message of supreme luck: the pilot nursing his ailing plane to escape over friendly land and not over the less friendly Channel; the crash in a populated area that resulted in minimal injury and no deaths; the pilot avoiding the misplaced wrath of the home guard by landing fortuitously in the convent courtyard. I believe this scene was something of a talisman of the luck I was to carry with me through the rest of my life.

The ties that bind—setting the stage

In Twelfth Night, Shakespeare remarked, "Some are born great, some achieve greatness, and some have greatness thrust upon 'em." If I have achieved any level of greatness, it has not been because I was born to it, although I may have received a happy genetic combination from my mother and father. My father's family were ordinary working folk. My paternal grandfather worked in the green groceries of Covent Garden, hauling bags of produce and working hard until he was 75. He carried bags of potatoes until he retired, then he died half a year later because he had no other interests in life. He seemed to be a diamond in the rough, and I remember that he was always very nice to me. For my birthday one year, he bought me a suit of clothes.

I also remember visiting my grandmother before and during the war. She lived in an older part of London called Lambeth. The toilet was outside in the back yard, and the bath was a galvanized tub filled with hot water from the stove. The house had no electricity, and gas lights brightened the night. I was too young to sense a dramatic difference in lifestyle, but my father was embarrassed by it. That was probably why we did not visit much.

My father started his work life sweeping the floors in a printing factory at the age of 14. During his career, he worked his way from that job to running the plant. Being the first within memory in his lineage to rise from labor into management, I am sure his life was stressful. He seldom, if ever, talked about his work at home. He seemed to keep everything inside, in his head, and was probably not a very good delegator. While openness and gregariousness did not seem to be strong elements of his life, perseverance and self-reliance certainly were, and he became general manager of the plant in his early 30s.

My mother's side of the equation was remarkably different. Her father was an artist and an antique dealer who spoke French and frequently took his young family to France on buying expeditions to stock his two shops. He had little personal influence on me as he died when I was only three years old, but I was to feel his gentle genetic influence throughout my life.

As a youth, I grew up in circumstances that encouraged me to expand my grasp to embrace all possibilities and to accept things the way they were, to work hard for what I wanted, and to make do with what I had. This influence of balance and the genetic disposition I was born with was to result in an ability to participate in the world around me and learn from it. Coupled with this was an almost instinctive quality that had people willing and wanting to work with me. I guess you could call it leadership. This leadership would take its time developing, however.

Head in the clouds

Aviation was also a strong influence in my early life. When I was four, I received a Christmas gift of a tricycle shaped like a biplane. That seemed to be the start of it. We lived in Edgeware, a suburb of greater London, when I was a boy. The nearby airfield in Hendon would hold air shows before the war, and, as a family, we would sit on the hillside and watch the planes circle below us. Biwing Gloucester Gladiators would zoom in, blasting barrage balloons out of the sky in a ball of flame, as Spitfires and Hurricanes put on a show. It made quite an impression on a young boy.

Just before the war started, a son in the family next door joined the Royal Air Force, and he used to talk to me about airplanes. This really lit the fire. I began by getting books about airplanes, and then I started drawing them and making models.

The day that World War II started, 3 September 1939, I was helping deliver milk. As an eight-year- old, I would go along with the milkman in his horse-drawn cart. I think I was even paid, although not very much. I just enjoyed being with the guy and delivering the milk. At about 10:00, he said, "Brian, you had better go home. There's something going to happen in about an hour." When I got home, Neville Chamberlain was announcing on the radio that we had declared war on Germany. No sooner had Chamberlain finished than the air raid sirens went off.

In anticipation of a German attack, the government had issued gas masks to everyone. There were special gas masks for children and even gas masks for babies that one could put the whole baby inside. My aunt was visiting at the time, and as soon as we heard the sirens, she blurted, "There's going to be a gas attack." My mother was really nervous about the possibility of such a thing, and so everybody panicked. My mother and aunt had heard that damp blankets stuffed around the doors were a good deterrent to keep the poison gas from seeping in. All of us— mother, aunt, sister, and I—crammed into a little cubby hole under the stairs with our gas masks on and damp blankets stuffed in the cracks, and we waited for the gas bombs to start dropping. After about half an hour, the all-clear sounded. It turned out that the presumed German attacker was actually a British plane ferrying the Duke of Windsor back to England from Germany.

Edgeware was a newly built section just northwest of London. Just prior to the beginning of the war, a number of Jewish people fleeing from what was happening in Germany were allowed to emigrate and settled there. In fact, the bus driver used to call my neighborhood Little Israel. We lived next door to Mr. and Mrs. Moses, which seemed strangely appropriate. As the war went on, none of these émigrés were allowed to join the armed forces, since they were still German citizens.

The war had little impact on me in its first year. I knew about the evacuation of Dunkirk, but the reality of being on a war footing did not come to me until I started seeing all of the young men going into the military. My father, because he was needed as one of the managers of his company, went into the war reserve and, because all of the young policemen had gone off to fight, the police auxiliary.

About a week after the incident with the Hurricane, I spotted a Bristol Blenheim overhead. Five black shapes emerged. Initially, I thought they were parachutes. When I realized they were bombs, I just held my breath. One exploded behind me: one exploded in front. A third went through the roof of a friend of mine's house. That one turned out to be a dud, but the shock of its impact flipped a knife out of a kitchen drawer and cut off his finger. He was only a kid like me—eight or nine years old—and there he was with a finger missing from a bomb dropped by one of our own planes. The Germans had captured the plane, and they had used it against us, which was not quite cricket. The war had become real.

Until air raid shelters in our school buildings could be built, it was considered too dangerous for us to go to school. Every other day a teacher would come to a house in the neighborhood where 5 to 10 kids of all ages would gather. She would give us lots of homework, and then move on to the next house. Some of my friends were evacuated to Canada or the Northwest, but my mother did not believe in evacuation. She also did not believe in air raid shelters.

Because the Underground was actually above ground in the outskirts of London where we lived, the government built large air raid shelters at the end of each road for families to go to during attacks. We never did. I would go to bed each night in the same bedroom I always used. When the bombing started, I could lie in bed and watch the searchlights with airplanes caught in them. Because I had become quite a spotter, I could usually tell which were German and which were British. Looking out my window, I could see the bombs dropping and the red glare of the explosions and the city on fire. Then the shrapnel from our antiaircraft shells would come clattering on our roof. In the morning, I would collect the bits of shrapnel to trade in school the way kids do with trading cards today. One day someone showed up with an unexploded incendiary bomb. We were terrified, but we looked anyway. Soon, the police came to take it—and him—off.

War was always with us. Our gas masks were our ever-present companions at school, and on many days classes would be held in the air raid shelter as bombing went on around us. In our neighborhoods, we played at war. Our section of London had had a lot of housing development going on just before the war, and there were big piles of bricks and other materials lying around just where the builders had left them to go to war. We kids made bunkers from the bricks and things and had our own play wars. One day, when I was supposed to be watching my sister, I simply brought her into our war zone. She must have looked up when she should have been ducking because an enemy rock hit her right on top of the head. It was only a little rock, but she was bleeding profusely. To make matters worse, she was wearing a white dress that was instantly splotched with red. I took her by the hand, whimpering home to mother, who nearly killed me at the sight of my bloodied charge. (It wasn't my fault. Maybe we shouldn't have been playing there, but everybody was doing it.)

My mother pretty much ran the family. She was used to taking charge. Her father, the antique dealer, was 65 when she was born and 67 when her sister was born. When my mother was 13, her own mother died, and she was put in charge of the household. The two girls went along with their 78-year-old father on his shopping expeditions to France because there was nobody at home to watch over them. They would sit in the railway station and talk with people while their father searched for deals on antiques. Back in Kensington, my mother helped run the shop. She was good with numbers, she knew antiques, and she understood how to wheel and deal. She also had a mind of her own. Her father vehemently opposed her marriage to my father. He considered that she would be marrying beneath herself, but she persisted. The ill will never went away between my father and grandfather, however. When my mother's father died in 1933 at the age of 85, my father refused to have any part of my mother's inheritance mixed with money he had earned.

By 1942, the British forces were really running short of men. My father was called to active duty as a staff sergeant in a field ambulance unit. He was in charge of such things as dispatching ambulances and accounting for bodies. This must have been a devastating assignment for such a quiet, internal man. He never spoke of his experiences, but he was a markedly different person when he finally returned home.

Sometime in 1943, the bombing came to a virtual halt. Then the rocket attacks began. The V-1 was a pulse jet that flew like an airplane. I would hear V-1s come rasping over the house on their way to somewhere. These things would eventually run out of fuel, fall out of the sky, and explode wherever they hit. Occasionally, I could hear the wind whistling through a V-1 gliding overhead with its motor off. I guess I was safe if I could hear it, but at the time I was not so sure.

We were never officially told about the V-2s, the first real ICBMs. If there happened to be a big explosion from a V-2, the local authorities would claim a gas main had blown up for some undetermined reason. I was told that you could hear a shock wave from the rocket before it hit the ground, but I never experienced this myself. People were not worried about rockets. They were far more worried about big bombs dropped with parachutes that sometimes left them tangled and dangling from the telephone wires.

By now, the supply of balsa had dried up, and I was making my model airplanes out of pine. I also started fabricating standard lamps and small table lamps for my friends' families, since there were none to be had in the shops.

On the playing fields

As a youth, I found school totally uninteresting. There was, of course, a war going on, but even so, my disinterest was far greater than that of my friends. What emerged from the confining frustration of school was a desire to compete in other ways, and I became something of an athlete.

Today's young people have athletics organized for them: leagues, eligibility rules, adult coaches, and referees. My youth had none of those things. If you wanted to play then, you provided your own organization. There were telephones, but not many. If you wanted a game of soccer or cricket, you got on your bike and went to other neighborhoods to find other teams. Then you would challenge them to a match and work out the logistics of time and place. It took someone to do that, and I decided that, because I was the one who wanted to play, I would have to organize it.

Competition on the playing fields and being a patrol leader in the Boy Scouts were enough for me until the 11-plus exams. This was the proficiency test for 11-year-olds that determined whether you had the potential for higher education. All my friends passed—and I did not. I was devastated. Had I failed as a youth taking that test today, someone would have probably diagnosed me as having a learning disability and prescribed some chemical cure until I conformed to his or her expectations. Then, it was up to me. Something clicked. I realized how important it was for me to lead, to be at the front, to win, and I took on academics as a new challenge, a new competition. From then on, I was never less than second, and usually at the top of my class.

As a result of my hard work, when I was 13, I earned a scholarship to a technical school. Actually, there was some question as to whether it was to be technical school or art school. I had become something of a budding young artist, progressing from flaming airplanes to landscapes and still lifes.

Weighing my options, however, I decided that the technical school made better economic sense. Besides, this technical school also had some very good sports teams. So, at the age of 13, I went to technical school, and although I did very well in school, I was also playing a lot.

After a few years of this, my mother was clearly not happy with me. "You never do any work," she said. "I can't understand how you do so well in school. It's about time you got a job." This was her way of announcing that she had arranged an interview for me at the deHavilland Engine Company at Stag Lane, Edgeware in Middlesex. I was upset by this turn of fortune. School and sports had become an easy and comfortable combination. It was a life that would be hard to leave. Nonetheless, there was no objection that could stand in the way of a mother like mine. I went to the interview—and at 16 became an apprentice at deHavilland's engine works.

Life as an apprentice

The job of an apprentice was to learn, essentially through a rapid-fire series of on-the-job-training experiences coupled with some formal training at the Hendon Technical School. (Technical school was two days and three nights a week in the first year and one day and three nights a week after that.) In actual practice, however, being a junior apprentice also meant being at the bottom rung of a seemingly vast hierarchy of experience.

At the time I joined deHavilland, the war had recently ended, but many of the workers were still the women who had joined the work force when the men were away fighting. Perhaps I expected gentler, more maternal treatment from women, but I remember to this day not only how good they were as workers but also how tough they could be on a young apprentice. I can only imagine that working in a

The engineer as a young man.

factory was a difficult transition for them from their lives before the war and that they wanted to steel these soft boys to the harsh realities of the world of manufacturing.

In many ways being an apprentice was like being an indentured servant except that I was paid 23 shillings and six pence each week (the equivalent of about four dollars in the United States at the time). Since a haircut or a movie was only a shilling then, I was doing pretty well for a 16-year-old, and I did not really mind the little abuse I had to take to get it.

During the first year, I learned how to file, chisel, drill, and mill—everything that goes on in a machine shop. Then I went to an area where we assembled and disassembled piston engines and deHavilland's first jet engines. Finally, I spent three or four months learning to draft.

In the second year, apprentices were placed in specific assignments. I worked for a while in a machine shop, then in a disassembly operation. The apprentices helped build a duplicate of the original Wright Flyer engine for the British Science Museum.

Years after their flight, the Wright brothers had had a dispute with the Smithsonian over which inventors were actually the first to fly. As a result of the ill will this generated, instead of donating their original Flyer to the Smithsonian, they sent it to the British Science Museum. Years passed, and the Wrights were recognized as the first to fly. Now the British Science Museum was preparing to send the original Flyer back to America, but they wanted a replica to stand in its place. I was part of the team at deHavilland that made the replica engine.

It amazed me to discover just how primitive aircraft engine technology was in the early 1900s. The carburetor used a wick that dripped gasoline directly into the intake manifold, where it vaporized and was drawn into the engine. The finished engine had so little horsepower that we simply clamped it to a workbench and started it right in the shop to test it. I felt even then that it was a marvel that the Wright Flyer flew at all.

As an apprentice, my first real connection with production engines was tearing down and overhauling Gipsy Minor, Gipsy Queen, and Gipsy Major engines. These were small, unsophisticated piston engines with lubricated cast-lead bearings. About the only real safety feature on these engines was a dual magneto system, so that if one magneto or one spark plug failed, the second would keep the airplane flying.

The Gipsy Minor powered the Tiger Moth. Tiger Moths are still flying. In fact, their owners meet at an annual convention. There were 27,000 Gipsy engines made, and they powered many aircraft around the world. One of them, the deHavilland Chipmunk, is used as a basic trainer even today. The Gipsy family of engines was one of the first to reach 1000 hours between overhauls—which, although it seems insignificant compared to today's jet engines, was quite a milestone.

Although these were marvelously resilient and durable engines, they were not much pleasure to work on, and I learned a lot about the relationship between engine design and engine maintenance. As an apprentice with relatively large hands, it was clear to me that the engineers who designed these engines must have known that they would never personally have to build them or take them apart. Some of the parts were held by bolts that each seemed to take half an hour of tedious, quarter-turn-at-a-time manipulation to remove.

At the time, all I could do was rant quietly about haphazard design practices, but the irritation of this frustrated apprentice stayed with the engineer I was to become. When I came to be a designer of parts myself and later as I oversaw the complex interaction of parts and components coming together as new engines, I always remembered that apprentice with large hands trying to turn tiny bolts in small crannies. As I look back, while I hated it then, I feel truly fortunate that as a lower-than-low apprentice I paid the price of someone else's engineering insensitivity. I vowed never to do that to anyone.

Twenty years later in my career, when I had the opportunity to oversee the reentry of General Electric (GE) in the commercial aircraft engine business, these early scraped knuckles and frustrations were redeemed. Those future engineering processes under my watch involved creating full-scale mockups and timing part removal and replacement. We set specific maintainability goals as part of the design criteria, and the design was not approved until we were satisfied that a mechanic in the field could maintain the engine with relative speed and ease. In fact, the maintainability of all of the engines that came from GE after the mid-1960s set a new standard for the attention we had paid to making them maintainable. Had I moved directly from engineering school to designing parts, I may never have learned—and passed on—this lesson.

The deHavilland engine plant at Stag Lane, Edgeware, was built at a converted airfield just outside London. Former hangars had become machine shops to produce the great volume of engines demanded by the war. Surprisingly, however, this was in no way a drab, dingy place. Skylights let in an enormous amount of light, and the place was generally clean and actually quite modern for the time. There was also a sense of family about the place. Every morning and afternoon, the teacart came around, and old hands and apprentices alike shared ritual breaks in the affairs of the day. In spite of the fact that as apprentices we were sent off to locate rubber hammers, left-handed wrenches, and other non-existent tools, the other workers really did appreciate us, and we learned a lot in the bargain, especially some very colorful swear words.

Craftsmanship

The machinists in the factory were, for the most part, self-taught craftsmen. They had started off as apprentices themselves and had learned by watching or by trial and error. Being a machinist at the time was something of an interpretive art, and the drawings were just a starting point for the old hands. As part of their craft, they would remove burrs or create other features that they thought would make the part better from a stress point of view. Our machinists automatically chamfered the parts they made, that is, instead of leaving a square edge on a part, they would add a 45-deg cut across that edge to slightly round it. The logic was that, if cracks were to form, they would usually start at the sharp edges where stress was greatest. By chamfering edges, the risk was lessened—and our machinists did that without any direction. As a result, the parts they made never looked exactly like the drawings.

We once received a few lots of parts made by an American supplier. The parts were made exactly to the specifications on the drawing. The people at deHavilland were disappointed with the work, feeling it lacked the craftsmanship they had come to expect from their own workers. On the other hand, we had to ask for exactly what we wanted if we were to get it from suppliers, another lesson I remembered. Still, I was impressed to have worked with people who so prided themselves on the quality of their craftsmanship. It felt good being around people like that, although I realize in retrospect how difficult quality might have been to control in an environment of personal standards.

Productivity was also hard to control. Although we used timecards, production machinists were paid on the basis of piecework. Whenever time checkers came around to see how many pieces should be expected out of a day's work, the machinists set their machines to run about 25% slower than normal. To attain impressive productivity improvements during a big crunch, all of these guys had to do was work at their normal rate, another lesson I learned about the culture of the factory.

In England at the time there were two classes of employment: factory workers and office workers. Disassembling and overhauling engines was hard, dirty work, and as a young man looking for something both cleaner and more mentally challenging, I set my sights on other assignments. Because I was doing well in school, at the age of 18, I was placed in the technical department. The first job I had there was in the foreign office, negotiating and coordinating the building of deHavilland jet engines in Sweden and Switzerland. While at the time I found this work a bit boring, especially compared to designing, I realize now how much it helped me learn to deal effectively with people who did not always share the same background and agenda.

As a surprising note, I discovered that the Swedes had an even greater pride of craftsmanship than we British did, and it was coupled with a much superior planning discipline. They were making Ghost and Goblin engines from deHavilland designs. (For some reason, deHavilland named their engines after spirits and sprites and those sorts of things.) These engines powered the Vampire and Venom fighters, respectively, and the Ghost was actually to be the first commercial jet engine, powering the Comet. The Ghost and Goblin engines that the Swedes were making—from the very same drawings and materials that we used—were better engines with much tighter tolerances. As a business, deHavilland never questioned this quality gap, perhaps because we did not want to hear the answer.

Designing ways

In the third year of my apprenticeship, I was assigned to the experimental engineer's office, where I got the chance to actually design aircraft engine parts. It is important to understand that, even though the jet age had begun, the height of technology was still the piston engine. Carburetor technology and spark timing had become extremely precise and exotic. To get more power from the bigger piston engines, they were supercharged, either with a direct-drive supercharger or some kind of turbocharger.

Superchargers compress the outside air before it goes into the combustion chamber. The more air, the more fuel it can hold and the more powerful the explosion. Supercharging was a practical necessity for high-altitude flights where the air is thin. A direct-drive supercharger draws its power from a belt, shaft, or gears attached directly to the engine crankshaft. A turbocharger uses the hot exhaust gases coming out of the engine to turn a turbine wheel. The turbine is connected by a shaft to the actual compressor. The turbo-supercharger was the conceptual forerunner of the jet engine.

In the state-of-the-art piston engines, complex combustion chamber and piston shapes were squeezing the most out of every ounce of no-knock high-octane fuel with very high compression ratios—and meticulously controlled detonation patterns. Critically designed crankshafts supported tremendous loads while keeping everything perfectly balanced.

By comparison, the centrifugal-flow jet engine was simple and did not demand technological precision. It was somewhat loosely fit together with plenty of slack to accommodate heat- and spin-generated dimensional changes in operation that we could not otherwise control. It achieved pressure ratios of about 3.5:1, that is, the compressor squeezed the air to 3.5 atmospheres of pressure before passing it on to the combustion chamber. (By comparison, the pressurization in today's engines,

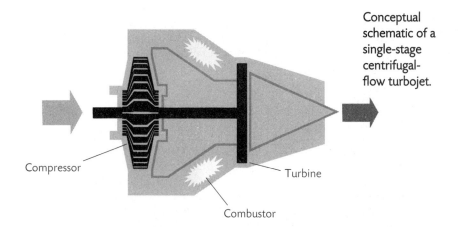

Conceptual schematic of a single-stage centrifugal-flow turbojet.

Compressor
Turbine
Combustor

such as the GE90, is nearly 40:1.) These new jet engines burned kerosene—lots of it and not very efficiently. Instead of an exotic carburetor with floats and endless valves, ventures, and governors as on the piston engines, they had an engine control that was essentially a block of aluminum with a couple of valves in it to control the fuel to nozzles in the combustor. The result was a specific fuel consumption of more than 1, that is, more than 1 lb of fuel burned per hour for each pound of thrust generated. (A pound of thrust is the force necessary to accelerate a mass of 1 lb 1 g—the force of gravity—32 ft/second/second.) Today's engines have specific fuel consumptions of about 0.3, that is, 3.3 pounds of thrust from each pound of fuel per hour.

As time went by, we were to achieve that fuel efficiency by, among other things, increasing both the pressure and the temperature of the gases rushing out of the engine. The temperature of the gases coming through the turbines of those early engines was between 1000 and 1200°F. The back-end temperature of today's engines is about 3000°F. The early engines proved that jet power was possible; however, they certainly did not prove it was efficient. In fact, there was even some question about them being technologically interesting.

Dr. Kerr Wilson, the head of deHavilland's stress office in the jet's early years, was unequivocal about their future. He said, "Those blasted Bunsen burners have no interesting technology for people like me, and they will never make it!" He was wrong, of course, but that is how it looked then, especially to Wilson, who as a doctor of physics and a mechanical designer, loved the intricacies of the piston engine. After all, a turbine engine was just one piece spinning around inside another piece. (Now, of course, turbine engines have gotten tremendously complex, and I am sure Wilson would have loved them.)

As an interesting footnote, our design process was hardly the vast enterprise of today involving thousands of people in thousands of cubicles using enormous investments in cutting-edge computer technology. In those days, the information processing technology simply was not available—we were still using hand-cranked calculators—but the way we approached design problems also seemed a lot more "hands-on."

Immediately after the war, motor car racing was still too expensive to gain widespread popularity, but motorcycle racing was comparatively affordable. As a result, as motorcycle racing became popular, motorcycle engines became cutting-edge technology. To help design the combustion chamber for a new small aircraft engine, the deHavilland design team bought a used Norton motorcycle, took apart the engine, and reverse-engineered that combustion chamber shape into our engine.

The Goblin

When it came to jet engines, our chief at deHavilland, Major Frank Halford, led the way for us. He and his right-hand man, John Brodie, were ultimately responsible for the Goblin single-stage centrifugal turbojet that entered service on the Vampire, a twin-boom fighter, in 1947. I am sure that Halford had been paying attention to what Sir Frank Whittle had been doing with the development of the jet engine, but testing these design concepts seemed more like a trip to the spare-parts room than the vision we have today of advanced technology in action. Turbo-superchargers already existed for piston engines, and so he just scrounged a compressor and, with a couple guys, did a stress analysis.

We kept the speeds low so that we could use an aluminum impeller—the spinning part that compressed the air—without fear of it flying apart. We made a diffuser case out of magnesium to keep the weight down. The combustor was individual combustion cans. The result was a simple single-stage turbine, and we had designed it and brought it to stress analysis with fewer than 20 people—fewer than 20 people to do all of the aerodynamics, thermodynamics, heat transfer, and stress analysis—and the result was an engine that ran quite successfully in a production airplane.

The Goblin produced about 3000 lb of thrust. By comparison to today's engines, it was awfully big to get that much thrust. Today, we generate comparable thrust with engines that take up about 7% of the volume of the Goblin. These early engines were also thirsty. The Goblin consumed well over 1 lb of fuel to create each pound of thrust. As I said, today's commercial engines are at least 300% more efficient.

The first Goblin I worked on, I destroyed. Actually, it was my job to cut a cross section out of an engine from front to back so that it could be used as a display model of the internal components. This taught me a tremendous amount about how this kind of engine was designed and manufactured, somewhat the way doctors learn about the workings of the body in autopsies.

The Ghost and the Comet

It was in this department that, as an apprentice, I also designed my first part for a jet engine, a small pickup tube that helped control fuel flow by measuring static air pressure at the inlet on the Ghost engine, the engine that was to power deHavilland's ill-fated Comet airliner. I moved from the experimental engineer's office before I saw my part actually go into production, but in my next assignment, the stress office, I was to be connected with the Ghost and the Comet again.

The Comet is an example of how a company with very limited resources took on a huge challenge and almost succeeded. The Comet was a beautiful, clean airplane with four 5000-lb-thrust Ghost engines buried in the wings where they attached to the fuselage. Instead of being riveted, the fuselage was bonded with a kind of epoxy glue material. Unfortunately, engineers at the time simply did not understand enough about fatigue in aluminum and stress concentrations. Each time the Comet flew to cruising altitude, the pressure inside the cabin stayed about the same as air pressure at 8000 ft, while outside the air at 30,000 ft was under much less pressure. To compensate, the fuselage expanded slightly. As the aircraft returned to earth, the pressures equalized, and the fuselage returned to its previous shape. The areas around the windows were not reinforced the way they are now, and they were subject to low-cycle fatigue, one cycle being every takeoff and landing. Eventually, the flexed metal lost its strength like a paper clip that has been bent back and forth too many times.

Sir Geoffrey deHavilland was still running the company when the Comet was introduced. He had started when he was about 20 years old and had been running the company for more than 30 years at that time. Sir Geoffrey was one of a select few people who had run an aircraft business based on first-hand knowledge of what would fly and what would not, and of what was dangerous to do and what was not. Perhaps it was this kind of confidence in personal experience that encouraged pressing beyond the analytic tools that were available to test concepts before they went into production. In truth, no one at the time understood—or could have been expected to understand—the implications of flying at very high altitudes and the effects of low-cycle fatigue on a bonded fuselage.

In any event, stress concentrations that began at the window corners propagated into the fuselage. The method of construction allowed the fuselage to crack, and the aircraft ultimately failed catastrophically. The first time this happened, a Comet was flying near the Italian shore and seemed to simply fall out of the sky. At first people suspected the engines.

A recovery team pulled the wreckage from the sea, and the first thing they noticed was that the turbine had ripped off each of the four engines. The investigators then suspected that the engines broke, causing the crash. Back at the engine company, we performed some tests in which we ran an engine at speed then released the rear mount causing the engine to pitch up abruptly. The gyroscopic force of the spinning turbine caused it to try to remain level. The result was that the turbine snapped off this engine just as it had on the crashed airplane, apparently, when the aircraft hit the water and rotated dramatically.

The engines were vindicated, but that did not help the deHavilland Aircraft Company. Later, tests in a special pressure tank proved that the fuselage had split in the area of a window. The window area was reinforced, and the Comet continued to fly, but deHavilland's fortune never did recover from this blow to its reputation. (DeHavilland soon was forced to merge with Hawker and later became part of British Aerospace.)

My fortunes were flourishing, however. Because of my academic record, after I was three years into my five-year apprenticeship, I was offered a state scholarship to attend the School of Engineering of Kings College, Durham University. This was quite an honor as there were very few of these scholarships awarded. After the initial euphoria, my practical side took charge. I decided to accept, but I deferred attending until after I completed the total five-year apprenticeship. This full apprenticeship would give me an advanced certificate in engineering (Higher National Certificate), the equivalent of a very practical bachelor's degree in mechanical engineering. I knew that whatever happened, I would always have that to fall back on. When I went to Kings College at Durham with my advanced certificate, I would then be working on the equivalent of an advanced degree in engineering. So, I pocketed my scholarship and stayed at deHavilland and continued to work in the technical office.

The Gyron

It was during this time that deHavilland began development of a huge axial-flow turbojet engine, the Gyron. Unlike centrifugal-flow engines, which compress incoming air by spinning it into an ever smaller volume, axial-flow compressors use blades to push the air into a smaller volume from the front toward the back of the engine, in the same axis as the spinning parts of the engine. As the air passes from the compressing effect of one set of blades, it is met by another set of blades that push it into a yet smaller volume. The air goes on like this from stage to stage, becoming more compressed until it is released into the combustor to be mixed with fuel and the mixture detonated. The pressure ratio of the Gyron was 6:1, almost double that of the Ghost or the Goblin. Increasing the pressure ratio was important for two reasons: it allowed the introduction of more fuel into a given volume of air, and the higher compression produced higher velocity gases shooting out the back of the engine. Compared with today's engines, a 6:1 compression ratio is still relatively low. The sheer size of the engine, however, would allow it to produce 30,000 lb of thrust—six times more than the Ghost—making it the biggest engine in the world in the early 1950s.

An axial-flow compressor is significantly more efficient than a centrifugal compressor, that is, it takes a lot less turbine power to turn an axial compressor to reach the same compression ratio, so that you would have considerably more effective high-speed gases coming out of the back of an axial-flow engine as thrust. What was surprising was that, while we were designing this monstrous axial-flow engine, our sales force was out telling potential customers that centrifugal-flow

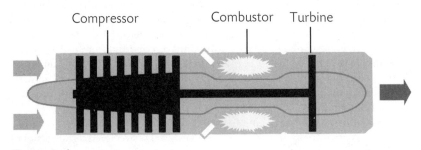

Conceptual schematic of an axial-flow turbojet.

engines were the only way to go. Nobody at deHavilland ever publicly admitted that we were working on an axial-flow engine.

To no one's surprise, the Gyron never found an application, but the Royal Navy needed a smaller engine for an aircraft called the Buccaneer. We spent days scaling everything down from the Gyron to create the Gyron Junior for this airplane. Ultimately, the engine was sold, and deHavilland had a production run of about 500 Gyron Juniors for the Buccaneer aircraft.

The university and beyond

Eventually, my five years of apprenticeship ended, and I headed north to Kings College in Newcastle. While much of the technical training was redundant, my work at the University of Durham filled in much of the theory that was missing from my technical school education. I noted, however, that most of the professors were really quite impractical. I had worked five years assembling and disassembling engines, designing parts, analyzing engine performance, and working side-by-side with recognized engineering experts. Now, I was listening to teachers who, with the exception of one or two, had very little comprehension of practical engineering. Embracing the academic approach was a bit of a challenge for me, but I reasoned that the honours degree would help me go after the higher paying jobs after graduation.

Attending the university did not end my relationship with deHavilland. I continued to work in stress and performance analysis during the five months of breaks and vacations in the university's school year. I also did a special study for the production manager of deHavilland's Levesden plant to put in a new production line. This brought me again to the realization that what we gained through craftsmanship, we were losing to poor planning. We had seen that the Swedes and the Swiss had excelled with our engine designs, largely because of the way they meticulously planned and controlled each step in the process. Still we had workers who resisted the planning process to allow themselves the latitude to do their work to the quality level, in the order and at the speed they chose.

Our machinists and assemblers were outstanding craftspeople. The truth is that these early jet engines really did not fit together very well, and we were continually reworking parts to make them fit. It is truly amazing that we ever had two engines that ran the same, but we had many successful engines coming out of the shops. We could have come closer to quality without rework, however, if our managers could have led our hourly work force to a greater appreciation of the benefits of better planning of the processes. While the project of laying out a production line for the Levesden plant was a great learning experience, resistance to change also made it a great frustration.

In 1953, Durham University awarded me an honours degree in engineering, and I returned to work full time at deHavilland. I now had a five-year apprenticeship behind me, the equivalent of an additional year-and-a-half experience in the design office and an advanced degree in engineering. I took the next natural step; I started looking for a better job.

I interviewed at a number of places. At Armstrong Siddeley, I met a Mr. Raymond, the developer of the Sapphire engine. At Bristol Siddeley, I talked with

Mr. Norton, the designer of the Norton motorcycle whose cylinder heads we had copied, and who was at Bristol at this time. Eventually, I received a number of offers, all at significantly better salaries.

I went back to the management at deHavilland, thanked them for a pleasant association, and told them that I planned to take a job at Bristol. "Well, Brian," they said, "you know you can't really do that." I was dumbfounded. "You're working on a very important program right now, and in the interest of national security, we have the option not to let you go." As I was about to vehemently protest the violation of my essential democratic freedoms, they added, "But we will match their salary offer."

I smiled quietly. I could continue to live at home, saving myself that expense had I gone to Bristol, and I could continue to work with people I knew—all with an impressive salary increase. Luck strikes again.

Making rockets

At deHavilland, I was now working in the rocket department. My first project was working as part of a team to design a rocket that could be used to assist the Comet in takeoffs from Johannesburg, South Africa. Johannesburg is hot and high. The air is thin as a result, and when you run it through the compressor of a jet engine, you get a lot less oxygen in the compressed air that feeds into the combustion chamber than you would at sea level. Fewer oxygen molecules mean less thrust from the engines.

Johannesburg's airport was at over 5000 ft. The Comet weighed more than 100,000 lb, and the four Ghost engines produced only 20,000 lb of thrust at sea level. This was enough for most airports but was marginal at Johannesburg's elevation considering the climb-out requirements to get over the surrounding hills.

Our solution was two Sprite rocket motors, rated at 5000 lb of thrust each. These rockets would run for two to three minutes and would effectively double the Comet's takeoff thrust.

We tested these rockets on stands at Hatfield. I went to visit the operation and was startled by an eerie observation. Every person who worked at the rocket motor test facility was a blond! Was every person working with rockets a Nordic Aryan? Instead of a conspiracy, the answer was chemically far simpler. The rocket operated by passing hydrogen peroxide through a catalyst yielding hydrogen and oxygen. Kerosene was mixed with this, and the result was ignited. It was the residual peroxide fumes that had bleached everybody's hair.

Another innovation of the rocket department was the Spectre, a variable thrust rocket motor. It was really a clever design that had plastic bearings and ran on hydrogen peroxide and kerosene in the same way that the Sprite did. The engine was being designed for a hybrid airplane called the SR53 being made by Saunders Roe. The plane was to be a fighter with a Viper turbojet engine to bring it into action and a Spectre variable-thrust rocket motor to provide additional power in combat. The plan was to sell this aircraft to the West German Air Force.

The Spectre engine worked very well. In fact, we ran it on an endurance test for more than 10 hours, which was an eternity at that time. Unfortunately, just after one of these long endurance tests, an air vice marshall showed up at the test facility and insisted that we restart the motor so that he could see it run. Instead of starting and running, it started and blew up, taking the test cell with it. We discovered that the seals had worn out, and redesigning the seals set us back quite a bit. It was work that we had to do, but a tap on the shoulder might have been a better way to find out.

After the redesign, we tested the engine on a flying test bed that was a converted Canberra bomber. The Canberra had two 5000-lb thrust jet engines. The Spectre produced 10,000 lb of thrust on its own. In addition, the higher the aircraft would fly, the more dramatic the Spectre's thrust would become. Jet engines have to suck their oxygen out of the air. As the plane goes higher and the air gets thinner, the thrust of a jet engine goes down. By comparison, a rocket brings its own oxygen. It will have the same thrust at 30,000 ft as it does at sea level.

The test pilot of our Canberra was a seat-of-the-pants goggles-and-white-scarf type of fellow who was definitely not impressed with our rocket motor. We told him to climb rapidly to 10,000 ft on the normal engines and then engage the rocket motor and continue the climb. We also told him to expect three times the normal thrust when the rocket motor kicked in.

He just laughed, shook his head, and said, "Sonny boy, don't worry about me. I can handle this."

We were all listening on the intercom as he reached 10,000 ft and engaged the rocket motor. "What the hell is happening here?" came the voice from above. "I think I'm going to heaven." That Canberra must have set a record getting to 20,000 ft for any airplane at the time. The flight test was certainly a success, but I think the pilot had decided that he would rather not fly it again.

Although the SR53 promised to give German fighter pilots an equal thrill, the German air force decided against it. Instead, they chose the Lockheed F-104 Starfighter powered by GE's J79 engine.

The Americans had been demonstrating a surprising ability to take ideas that had begun in Europe, develop them, and bring them to market faster than the Europeans. I remember that when we were working on the Gyron Junior, management brought in a GE J47 engine for us to examine. We were amazed. The J47 produced lower thrust than the Gyron Junior, and it had a different combustor, but other than that it was technologically strikingly similar to what we were designing. This was an engine that, at the time we were designing the Gyron, was already in full production, flying on the Boeing B-47 Stratojet bomber and the North American F-86 Sabre Jet fighter.

By the time the autumn of 1956 rolled around, I had been working at deHavilland for nine years. Growth seemed nonexistent. People I knew were doing the same jobs they had been doing when I was an apprentice, and it seemed that this would be a good time to get some new experience. I considered in my optimism that I could always come back to deHavilland and never miss a beat.

An advertisement in *Flight* or the London *Times* or the Manchester *Guardian* spoke of a jet engine company in the American Midwest that was looking for engineers. Intrigued, I applied.

CHAPTER 2
Coming to America— and to General Electric

I had guessed the unnamed midwestern American aircraft engine company I had applied to would be Westinghouse. I was wrong. Soon I found myself talking with two gentlemen from General Electric's Aircraft Engine Group at the Dorchester hotel in London. Gunther Diedrich was a continental German expatriate who had been an engineer in the Heinkel jet engine operation during the war. Now he was both part of and a facilitator for the great brain drain of European technical talent to America. The other man on the interviewing team was Wally Dodge. Dodge was what we Brits thought of as typically American. He had a close-cropped burr haircut and wore an attention-getting window-pane-plaid suit with a bright bow tie. He spoke loudly and had strong opinions he was quite willing to share. We hit it off right away.

Dodge and Diedrich offered me a job at GE's aircraft engine plant in Evendale, Ohio, just outside Cincinnati. The Evendale plant was a huge complex that GE had acquired in 1949 to help fill the demand for J47 engines. By 1956, it had also become an engineering center of some note. They offered me $6000 a year, which seemed like a small fortune. In retrospect, the future of deHavilland definitely was bleak, but that did not really enter into my thoughts at the time. After my brief job search following graduation, I concluded that opportunities were plentiful. GE seemed like a place where I could learn something. Besides, my wife, Jill, really wanted to go to America. I accepted their offer without even visiting first. Another piece of luck was finding that Cincinnati was a great place to live.

Our foreign adventure

America was planned as a two-year adventure for us, and Jill and I promised our parents we would dutifully return. They were worried because Jill was now pregnant with our first child, and they had heard that medical treatment in the United States was below the standards of national health in England. It was with great parental concern that they, along with our whole family of aunts and uncles and nieces and nephews, bid us farewell at Heathrow Airport on a blustery February morning in 1957. As we were about to depart, I lifted a favorite niece to say good-bye and sprained my back.

After a stop in Shannon, Ireland (where I visited a doctor for my back pain and purchased the most expensive aspirin I have ever had in my life), our plane arrived in Detroit at three o'clock the next morning. The weather was bitterly cold; ice and snow covered the ground. One of the first sights I remember was a burly policeman with a pistol strapped to his waist. Policemen never wore guns in England. What kind of place had we come to?

From Detroit, we flew to Chicago to stay with an aunt for a few days. Chicago had just had a huge snowstorm, and snow was piled higher than I had ever seen snow before. The temperature was –10°F. In my aunt's apartment, it was about 90°F, however. After five days of this, even the Greyhound bus trip to Cincinnati was something of a relief—but only something. There are hardly any two places in the whole of England that are as far apart as Chicago and Cincinnati. It seemed an interminable journey through the barren winter farmland of Indiana.

A fellow named Erwin Snitzer, who was to be my immediate boss, met us at the terminal. He took us to the Sheraton-Gibson hotel in downtown Cincinnati, where Jill promptly became ill in the elevator. (Much to my surprise, the cleanup service appeared on our bill—another interesting welcome to the United States.)

When we arrived in Cincinnati, we had $200 to carry us until I started receiving my salary. Snitzer quickly connected us with a real estate agent who found us an apartment relatively close to the plant, and within walking distance of shopping, in Swifton Village. We put down $120 to close the deal and spent another $20 on food, leaving us with 60 bucks to our name until I would get my first check at the end of the month.

I was used to easy bus or train travel to work. In fact, in England, I could easily walk to work if I wanted. While our apartment in Swifton Village was close to the plant as the crow flies, Cincinnati's bus routes had nothing in common with birds. To go to work by bus, I would have to take one bus downtown and then another out to the suburb of Lockland where I worked in GE's Alien Office. Although I lived about five miles from work, the bus ride to get there was about 25 miles. Clearly, I needed a car.

I had heard all of these stories about unscrupulous American used car salesmen, and I was determined I would not be taken. I would buy a new car. At the local Ford dealer, I test drove one of their demonstrator models, a Fairlane with one of those big V-8s used by police interceptors. I was impressed. When I told the salesman my financial situation, he steered the conversation to used cars, but I was adamant. We eventually settled on a blue and white two-door, with a six-cylinder engine, a column-mounted straight-stick transmission, and rubber floor mats for $2350. Two fellow occupants of the Alien Office, whom I had met only three or four days before, agreed to cosign a note for me, and the new Ford was mine. We had been in Cincinnati less than two weeks. Already, we had leased an apartment and were driving a new car, and I had yet to receive my first paycheck. This was truly going to be a great adventure.

The Alien Office

GE's customers for jet engines were military, and national security demanded that foreigners like myself be thoroughly cleared before they had any contact with GE jet engine technology. Besides Englishmen, there were Australians, South Africans, and a number of Germans—even some fellows who had worked closely with Werner von Braun—all waiting for security clearance. I thought that since we British were allies of the Americans during the war, the clearance process should go fairly quickly for us. Surprisingly, it was the Germans who got their clearances well before any of us did. They, of course, had gone through a strenuous preliminary security check even to get in the country, but we allies were still miffed as we saw the Germans heading into the main factory while we cooled our heels in the alien facility.

As I sat and waited, I developed something of an aura of seniority in the Alien Office. (Sort of like being the head prisoner in a prisoner of war camp, this was not an honor one generally seeks but is willing to accept, all things considered.) Even though I was younger than most of the people in the office, I had been through an extensive apprenticeship, and I knew the jet engine business quite well. I also spoke English better than the Germans. Again, I was having leadership thrust upon me, and I found myself standing up for my fellow aliens as we attempted to find something meaningful to do while we waited. The process of waiting for a clearance took almost a year for me, and while that was frustrating, I gained invaluable experience in dealing with my fellow foreign nationals, experience that otherwise I might never have gotten.

Finally, the clearance came, and I was allowed to go into the plant, assigned to a German expatriate named Peter Kappus. I continued work on a lift-fan design for a vertical takeoff aircraft that I had begun in the Alien Office, but now, even though the technology we were working with was unclassified, I was officially cleared.

Working at GE was remarkably different from working at deHavilland in many ways. For one thing, all of the people running the various programs were remarkably young. GE had not even been in the aircraft engine business at the beginning of World War II. In fact, the first aircraft engine they built in 1942 was also America's first jet engine. Although GE made countless turbo-superchargers for B-17s and other aircraft during the war, they never made a piston engine. Becuase there were no traditional solutions to the problems people would face in designing these totally new jet engines, there was no cadre of wise old experts yet. As a result, it was a young engineer's paradise.

Another difference was that GE had the resources to support cutting-edge design efforts. At deHavilland, the technical office had a few people in it. They had just gotten an electronic calculator in the last year of my apprenticeship, and it was hardly world class. At GE, we had the best electronic calculators available, and technical offices were staffed with hundreds of engineers. Clearly, GE was able to cover some very large bets and was willing to take some chances.

Lift fans

Lift-fan technology was one of those chances. Efficient vertical takeoff and landing (VTOL) has always been a dream in the aviation industry. It turns out that the weight and the technical complexity of VTOL aircraft make it cheaper to build runways than to manufacture and operate these airplanes. On top of this, even after this kind of airplane is in horizontal flight, the additional weight and the aerodynamic complexity compromises the craft's performance or combat mission compared to a standard design. The only customer who could afford VTOL aircraft was the military, and then only for specialized, limited use. So, we were really going after a specialized market with our lift fan.

The airplane we were designing this engine for was called the XV-5A, and our customer was the U.S. Army. Our design called for a huge fan in each wing. The fans were covered with louvered slats on the bottom of the wing and doors on the top that opened for vertical takeoff or landing. After the plane was in the air, the slats and doors closed for horizontal flight. The fans were tip driven, sort of the way a water wheel works. Two of GE's J85 engines provided a fast, hot flow of gas. This flow was then diverted to the tips of the blades to set the fans spinning. It was quite an elegant design, and it worked. The engine was able to push 13 times more air through the lift fans than was going through the J85 gas generators, more than enough to lift the XV-5A fully loaded. After the wing-mounted lift fans raised the XV-5A high enough vertically, diverter valves would shift the flow of hot gas to the rear of the aircraft for normal horizontal flight.

We demonstrated the performance of this engine-fan system on a test stand at our facility in Peebles, Ohio, and the army awarded us a contract to build two airplanes—not two engines, but two testable aircraft—for $12 million—a minuscule sum by today's standards, and not really very much then either.

I think it was one of the most unusual programs in aviation history. The engine manufacturer, GE, was the army's prime contractor for the airframe. We developed aerodynamic models and wing designs in connection with North American Aviation. We then subcontracted with Ryan Aeronautical to develop the airframe who further subcontracted with Republic Aviation to conduct the flight tests, which were completed at Edwards Air Force Base. NASA was also involved in the process by advancing and testing technology. In spite of this complexity, we did manage to build two airplanes and flew both of them in a very successful program.

Pete Kappus had started working on this lift fan idea sometime in 1955. I was involved in the preliminary design work from the time I arrived at GE in 1957. Although Pete had come up with the idea of a lift fan, the demonstration program was headed by Art Adamson, a very creative team leader. The system manager was Ted Stirgwolt. Both of these men were outstanding mentors, and I remained friends with them throughout my career at GE. By late 1959, our demonstrator engine was being tested, and by 1962 we had an airplane flying. This was a great learning experience for me as I was involved in all aspects of the program from product development through design and then into the testing of the system. It enabled me to learn much about how GE worked at all levels. It also gave me extensive experience in working with the government and with outside vendors and partners. What really impressed me was how much decision-making latitude a low-level engineer at GE was given on all aspects of a program.

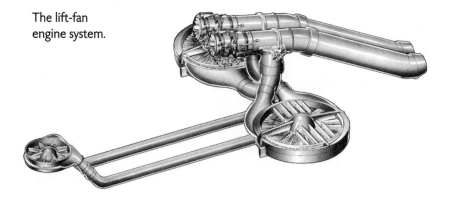

The lift-fan engine system.

Ryan XV-5A hovering.

The XV-5A performed remarkably well at GE's flight test facility at Edwards Air Force Base. In the first attempt to lift vertically, the plane slowly rose several hundred feet, hovered in midair, then slowly settled back to earth. After a few months of practicing vertical takeoffs and transitions into horizontal flight, the test pilot had become amazingly confident in the XV-5A. One day, without notice and certainly without permission, he took off vertically, hovered at about 30 ft, rotated the ducting for normal flight, and shot off into a barrel roll—a complete 360-deg rotation—over the California desert. Neither the army nor the GE team was amused, but the pilot had proved the plane could perform.

The GE1

While all of this lift-fan/cruise-fan work was going on, I was also part of a task force to develop an entirely new engine for GE. Gerhard Neumann, who was now leading GE's aircraft engine business, considered that our mainstay product, the J79 engine, was technologically out of gas and that we needed a new engine to replace it. Pete Kappus did a preliminary study, and then Fred MacFee was brought from our sister plant in Lynn, Massachusetts, to lead the design task force. My job was to head the team designing the compressor for this new engine. I had never really designed any compressors as such before, and so this was to be quite a challenge.

MacFee had just come from heading the J85 program in Lynn. Although the Lynn plant near Boston and the Evendale plant near Cincinnati were technically part of the same business, they often acted more like competitors, and the sibling rivalry between these sister plants could be surprisingly dysfunctional. Gerhard Neumann had spent most of his GE career in Lynn until that time. He had a professed love for Boston and his sailboat, and he did not care much for Cincinnati. When we transferred the headquarters of the aircraft engine business to Evendale, the engineering capital also moved there.

When the exodus began, Eddie Woll, a top manager in Small Aircraft Engines in Lynn, refused to come. He pulled some people together who shared his vision of keeping the Lynn River Works going and said, "You build those big engines in

Evendale, and we'll continue to build the small engines—T58s and J85s—here." By being a bit confrontational and outspoken, Woll and his crew in effect saved Lynn. That salvation also solidified a subtle distrust that bordered on a siege mentality in Lynn at times. The truth was that, although Neumann ran the business, Woll now ran Lynn. Neumann would try to break through that, but he never really succeeded. In years to come, even though Neumann had moved the headquarters back to Lynn, when he was there, he was in Eddie Woll's territory. Like two continents on the surface of the Earth, Lynn and Evendale not only had an ocean between them, but they were slowly drawing farther apart. This separation allowed them to continue to develop independent cultures, independent visions, and independent goals even while being part of the same business.

In the process of working on the compressor design for the GE1, I visited Lynn several times. In fact, the drum compressor design that our team decided on was ultimately derived from technology developed in Lynn's small engine operation. I cannot say that at that time I wanted to work there. The ride from the airport to the plant passed through the dregs of former industrial glory that was slowly passing into decay. GE's preferred lodging at the time was a hotel in Lynn that I could only describe as a real dump. I was quite happy after a visit to leave the Lynn plant to the people who saved it.

Moving to Lynn

Back in Evendale, I was genuinely enjoying my work on the GE1. This high-profile project was bringing me into close contact with some of the best engineering talent in the business. As much as I enjoyed the intensity, however, it was becoming increasingly clear that I could not focus my full energy and attention on both lift fans and the development of the GE1 compressor. Then I received a call from Dave Cochran, who headed the Flight Propulsion Development department, ending the dilemma. "Brian," he said, "we decided that you're going to move to Lynn." I thought of that drive through north Boston to Lynn and the dismal hotel, and I wondered how I would tell Jill.

After five years in Cincinnati, we had three little children, we had just moved into a new house that we had built in Greenhills, we were feeling somewhat settled—and now I had to tell Jill we were moving to a place I could hardly describe as a step forward from my initial impressions. I was not very enthusiastic about my new job either, heading the lift-fan project to bring it into production. I had great doubts that it was ready to be sold.

In June of 1962, without much enthusiasm, we moved to Lynn. Unlike the car ride to the plant, the area turned out to be beautiful. We found a very nice home in a place called Boxford, a community full of transplants like ourselves who were warm and friendly. A lot of the people we met there are still close friends. So, from the family standpoint, the move worked out well. Now I had to work on marketing the lift fan.

Our XV-5A was comparable to the Hawker-Siddeley Harrier. The Harrier had swiveling nozzles that turned to vector the thrust of a Bristol Olympus engine from vertical for takeoff to horizontal for normal flight. This system was heavier, less fuel efficient, and did not have the capacity that our lift fan system had, but it did have the support of the U.S. and British governments. In fact the U.S. Navy and Marine Corps invested well over $200 million in the development of the Harrier. Our budget from the army, on the other hand, did not allow us to iron out all of the bugs much less fly our XV-5A to air shows to generate support and interest.

We soon had two XV-5As flying in our test program. Unfortunately, we lost one of the aircraft early in the program when someone released a lifting harness with no weight on it that was attached to the hovering aircraft. The harness flipped over the wing and was ingested by one of the lift fans. The airplane went out of control and crashed, taking the pilot, and the program, with it.

The Harrier ultimately won the VTOL battle, but we pressed on. Trying to save our investment in lift-fan technology, we decided to rotate the lift fans 90 deg and use them as cruise fans. We even had a preliminary design that we had worked out with Grumman to put these engines on their Mohawk, a twin-turboprop observation plane and gun platform. The performance on paper looked spectacular, but this would definitely be an ugly bird. The idea never made it past the preliminary design phase.

The lift-fan project was slowly fading away for me. Its impact was to change the world of aviation in the future, however.

The legacy of the lift fan

Our work had demonstrated that we could move huge amounts of air with lift fans. When we turned them 90 deg, we had marvelously efficient cruise fans. GE even designed and built a demonstrator cruise fan with a big J79 engine just to show how much thrust we could produce. The logical leap to a high-bypass turbofan engine was not far behind. The air force had wanted just such an engine to power a huge troop- and cargo-carrying aircraft they were thinking about to make ground forces more air transportable.

In terms of core engine or gas generator technology, the GE1 was proving to be the formidable building block that Neumann wanted for the future. With such an efficient core design as the GE1, we could add a fan or an afterburner or a thrust-vectoring device to satisfy any number of future applications. MacFee and the team had identified 30 potential engine designs using the GE1 core, engines that could power seven totally new aircraft types.

The air force was interested. Don Berkey headed a GE team that set to work designing an engine that met the giant transport need. They settled on an 8:1 bypass ratio engine using the GE1 core and a fan that had one set of full-sized blades and one set of half-sized blades. The paper design then became reality as the GE1/6 demonstrator engine.

The successful demonstrator run convinced the air force that we had made the high-bypass breakthrough, and the first production engine to come from the GE1 building block concept was the world's first high-bypass production engine, the TF39, which I will say more about later. I was lucky enough to have been part of both the cruise-fan technology and the GE1 building-block venture that paved the way for this historic engine—one that was to be the design cornerstone for all future commercial aviation propulsion.

Adopted by the East Coast establishment

Meanwhile, I was still in Lynn trying to market our existing lift-fan idea. After a year and a half of frustration flying all over the country to sell this concept, I went to Eddie Woll and said, "Look, this project is not going anywhere. I want to go back to Cincinnati." To my surprise, a miraculous transformation had occurred without my knowing it. After a year and a half, I was now seen as a Lynn person. Woll could not understand why I would want to defect to Evendale, and he promised that I would find much better opportunities by remaining at Lynn.

"We'll put you in charge of advanced design," said Woll. "You can help Marty Hemsworth."

Hemsworth was one of the best engineers who ever worked for GE. He was an easy-going, good-natured guy, but he sometimes would get so intrigued with the details of a problem that the total picture seemed to fade in importance. He was a great engineer, however; and there is no doubt that he was a superb technical leader for GE over many years.

I worked with him on a T64 gas regenerator for a while, but soon went back to Woll. "Eddie, this is not working," I said.

"I'll put you in charge of Advanced Projects," said Woll.

"What in the world is Advanced Projects?" I asked.

"Well, you know, we're going to do this, and we're going to do that." I was now a confirmed Lynn person in his eyes. Eddie clearly wanted me to stay, and there would be no getting away.

About this time, Art Adamson had transferred to Lynn to head the J85 project, one of our biggest production engines at the time. From my perspective, Art was probably the most intelligent person working at Aircraft Engines and another great engineer. Art had been something of a mentor for me when I was at Evendale, and I genuinely enjoyed working with him. Since I really did not have that much to do in Advanced Projects, I spent a lot of time with Art on the J85. The engine had a lot of problems at the time, and I was continually calling meetings and getting things resolved, all without being formally connected to the project.

Eddie Woll, who ran everything at Lynn, was also a bright guy, but for some reason, he could not abide Art. As soon as Art would go out of town on business, Eddie would call me. "Brian," he would say, "Art's going to be out of town for a week or two. I want you to run that damn project for me."

"I can't do that," I would object. "Art's in charge, not me. I'm not even part of that project."

"I don't care," Woll would respond. "That's what I want you to do while he's gone."

So, I would call Art and ask him what he wanted done, and handle the meetings and run the project more or less the way he wanted. I think Art and I worked everything out without Eddie really knowing we were doing it. This whole period in Advanced Projects taught me some very valuable lessons about managing people who did not work for me. Enthusiasm—a real passion for the work—carries a lot of weight. If you care enough to take charge, and have some reasonable ideas, people follow.

CHAPTER 3
Earning My Stripes— Leading Engineers

The year was 1963. I had been working in Advanced Projects at Lynn for a while, and I was feeling rather overlooked. Then the golden opportunity Eddie Woll implied would occur surfaced in the form of the CF700. The CF700 was another proposed derivative of the J85 engine, one that had been conceived in the late 1950s when GE began a thrust toward more emphasis on the commercial engine business. The CF700 was essentially a J85 with a low-pressure turbine and fan on the aft end of the engine. (This was the same configuration we had used on the CJ805-23, GE's J79 derivative that powered the largely unsuccessful Convair 990 airliner.)

The CF700 was intended to power aircraft in what was expected to be a relatively small niche market, executive and business jets. Ralph Cordiner, the Chairman of GE at the time, and Juan Trippe, the CEO of Pan American Airlines, had decided that a great market awaited a business jet powered by a turbofan, and that the CF700 was the engine to do it. As the engineering design manager of the project, I was charged with making that dream a reality. Although this was a relatively small project with a relatively small team, the chairman's interest gave it a disproportionately high profile.

Initially, we planned to put the engine on a deHavilland 125, which was being powered by a Viper engine at the time. By comparison to the Viper, the CF700 would reduce fuel consumption by 30%, enough to make the new engine worth the effort. I flew to England to meet the deHavilland 125 project team, and it turned out that some of us had been apprentices together.

We started working toward getting the engine on that airplane, and things were progressing well when a Mr. Wilkens, the chief engineer at deHavilland, came to me and said, "Brian, this is not going to work. We'd rather you put your engine somewhere else. We've got an airplane that's flying and we don't need this help." It turned out that he and John Borger, the Pan American project manager,

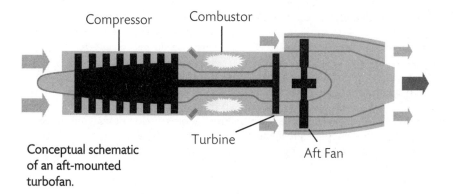

Conceptual schematic of an aft-mounted turbofan.

had just had an altercation. Borger came in a few minutes later, and he was still complaining about Wilkens and going on about how this was his project. "Now, John," I said. "You've just got to learn how to work with these people." Borger was a great engineer, and his work made Pan Am the world's first international airline, but he could certainly be difficult to work with. Unfortunately, the British were unwilling to accommodate his style, and they missed out on the sale of 600 airplanes. We abandoned the idea of the CF700 on the deHavilland 125 and got on a plane and flew to Paris.

A French connection for the CF700

In Paris, we met with Paul Chassagne, manager of Dassault's Mystere 20 program, and Bernard Larudier, the director of propulsion. The Mystere 20 was already flying with Pratt & Whitney JT12 engines that worked well enough but had high fuel consumption. We had brought along our own aerodynamics person, John Kutney, who had designed the nacelle for our first commercial fan jet, the CJ805-23. Together we designed a horseshoe inlet for the back end of the aircraft that would be renamed the Falcon 20. We did it to save weight, which it did, but it certainly was an ugly design. Marcel Dassault took one look at it and said that this odd thing hanging on the back would never work. We assured him that it would be great.

As a result of our salesmanship, we won the contract to put the CF700 on the Falcon, but actually selling the airplane was not going to be easy. The first problem was that GE had very little credibility in the engineering ranks at Pan American, the people who would be doing the actual marketing. This had something to do with a reputation developed when GE was supplying turbo-superchargers for Boeing's Stratocruiser that was being operated by Pan Am and Northwest. GE decided to terminate the supercharger operation and advised the airlines to buy all our spares immediately as we were no longer going to be supporting them. Needless to say, this left them feeling seduced and abandoned. In any event, the Pan Am people all loved Pratt & Whitney and had little time for GE. Since they would be responsible for selling the Falcon with our engines, we had a lot of work ahead of us to win them over.

One of the first things we did was to visit their service shops to get a good feeling for what it took to maintain an engine. That was an important experience. It brought me back to my days as an apprentice taking apart piston engines and wondering what kind of engineer would design something so hard to disassemble and reassemble.

Back at the plant, I learned that it was absolutely essential for an engineer to walk the shop, seeing how his parts were coming. Not only did that produce a real sense of satisfaction to see the hardware go from drawing to reality, but also engineers close to problems could help fix them. The other value in walking the shop was to verify that people were actually working on your project and not working on something else, but charging their time to you.

Eventually, we put our CF700 on the Falcon with the horseshoe inlet—and the aircraft did not perform anywhere near as well as we said it would. In fact, it did not even perform as well as it had with the old engines. The problem was not the engines

Dassault Falcon 20 with GE CF700 engines.

themselves but the nacelle. The design we came up with was causing so much drag that it was forcing the engines to work too hard and, consequently, to burn too much fuel. We got John Kutney back along with Arnie Brooks and some of the people from Dassault to redesign the nacelles, and we eventually got the drag down. We met the performance we had promised, but it is always embarrassing to have to fix things, especially after someone important has told you your design would never work.

After we had resolved all of the problems and the propulsion system was performing as we expected, we began working with our GE team on a business plan for the project. I said that I thought we would sell 300–400 of these airplanes, and everybody had a good laugh. They figured we would be lucky if we sold 50. After all, who would want all of these biz jets? It turned out we were all wrong, and there are now thousands of these airplanes out there.

The CF700 was my first real managerial role, and it was a success. Not only did I manage the engineering end and interact with many people around the world, but because Fred Garry, the overall project manager, came to the job after we had already started, it had been up to me to make many of the early decisions. It reminded me of my role in organizing sports contests as a youth. If you wanted to play, you had to fill the vacuum by getting on your bike and finding a game. And I was always willing and enthusiastic about doing that.

I also learned a lot about people. I grew up believing that if you built the best product, it would just sell itself. As a young engineer, I thought sales and marketing people were useless. In reality, it turned out that the most important people in any enterprise were the ones who sold the product. Of course, they had to have a good product to sell, but the people selling were the ones who could make the difference between the success and failure of a perfectly good product. And in selling—it turned out—relationships counted almost as much as the product itself. This was quite a revelation to an engineer.

Lynn Production Engineering

Although all I had wanted to do on the CF700 was to get the job done, someone up there must have been watching. In late 1965, Gerhard Neumann and the executive team decided to reorganize the business. There was no longer to be a Small Aircraft Engine Department. The business was going to be divided between military engines and commercial engines. Engineering, manufacturing, and quality would be functional, reporting through their independent chains of command to separate vice presidents or general managers. Fred MacFee was to be the head of all engineering: service engineering, development engineering, advanced engineering, and production engineering. Production engineering for the business was to be under Fred Garry—and I was to head Lynn Production Engineering.

I was delighted to get the job. This was a huge promotion and something I had never aspired to. I was 34 years old, and many of the people now working for me were much older and more experienced than I was. (I also found, after I got in the job, that they were making more money than I was as well, but I considered that wage parity was something I could look forward to.)

I took over the job from Sam Levine, a man I considered one of the best managers I had ever met. He was a gruff guy, smoking these miserable pipes all of the time, and he really knew how to let you have it if he felt you did something to deserve his ire, but he was an outstanding team builder. Sam had hemorrhoids the last few months he was in the job, and I found that, after I was sitting in his chair for a few months, I started getting the same symptoms. Unlike Sam, I got rid of the chair, and the problem went away. Maybe Sam had been using the chair just to give himself that gruff, bull-of-the-woods edge.

This was an exciting period at Lynn. Engineers love problems—they see them as opportunities—and we had plenty of opportunities. GE's small engine lines are marvels. They have stayed competitive over the years even though the initial designs may be fairly old, proving that, if you have fundamentally good designs, they can last a long time. The reason these engines could do this is that we continually applied new technology, new materials, new methods, or new design techniques whenever we could make the engines better and more sellable by doing it. When I took over as production engineering manager, the T58 line, which had begun production in 1956, was still going strong, and it continued that way until 1984. (T58s are still flying all over the world, and GE is still making good money selling spare parts for them.)

In 1965, we had a derivative version of the T58 that took it over 2400 shp and a version of the T64 that produced nearly 5000 hp. The J85 and the T64 were also in the middle of growth programs, and we were busy certifying the T64 on the H-53, a huge navy helicopter.

Because we were doing so much work for the navy, that branch performed the oversight of our government contracts at Lynn. They had a tough program quality manager for the T64 named Jack Horan. We called him no-crack Horan because he insisted that these development engines that we were struggling just to keep running come through a 150-hour test regime with no crack anywhere in the engine, an almost impossible task. This program was Eddie Woll's baby, and Horan was one of his drinking buddies. Woll thought he had a great relationship with Horan, but we never saw any relief at the lower levels. Ultimately, we learned how to make outstandingly durable engines that satisfied even no-crack Horan, and so I can say that he did a great job for the navy. The T64 has proven itself to be an outstanding engine. It is still on a three-engine version of the H-53, and it

Earning My Stripes—Leading Engineers

will probably be around for another 10 or 15 years because there is nothing even on the drawing boards to replace it.

The J85 was just going into service on the F-5 Freedom Fighter. And, of course, the Vietnam conflict was still raging. We had a number of helicopter engines and J85s involved there, and developing field repair strategies was a constant challenge. We sent a number of people to Vietnam to look at the kind of problems the military was having maintaining engines in the dirt and the muck of combat. Gerhard himself took a trip to Vietnam and visited the front lines and saw what terrible conditions the mechanics were working under to fix the planes and engines. He came back resolved to make it easier for the front line mechanics. Gerhard insisted that we restructure our maintenance procedures to what could be done in the actual conditions of war. One innovation he thought up during his visit to Vietnam was to create display boards of engine parts showing which ones were serviceable and which were not. Our military customers at the field maintenance levels certainly appreciated that kind of help—although I am not sure the higher ups were all that pleased—and our experience there taught us more about listening to the actual person using and maintaining our products.

The army was looking for a more efficient helicopter engine to replace the current ones on their Hueys. We had just completed the design of the GE12 demonstrator, which would become the highly successful T700, for this purpose. Art Adamson, who had been managing the J85 program, became the proposal manager for the GE12, and the relationship that Art built with the army helped convince them that this was the engine they needed. The GE12 was something of a breakthrough design for us in that it was an axicentrifugal flow engine. The first few stages of the compressor were axial, meaning that, as the outside air was being compressed, it moved parallel to the centerline, or axis, of the engine, or from the front toward the back of the engine. Then the air went into the centrifugal part of the compressor. Here, it was compressed further as it was spun from the center toward the outer edges of the engine. Although axial compressors were the current state of

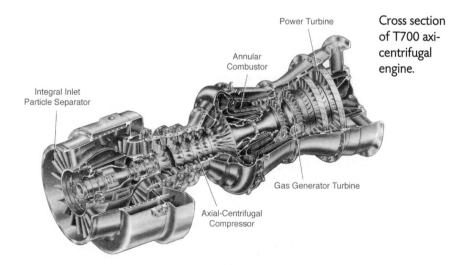

Cross section of T700 axi-centrifugal engine.

the art, GE had a number of engineers skilled in designing centrifugal compressors. The first jet engines developed in the United States had used centrifugal compressors, and some very smart and very young GE engineers designed them. Many of these engineers who had worked on engines during or shortly after World War II were still at Lynn and were only in their late thirties or early forties.

At about the same time we began work on the GE12, we started on the TF34. The TF34 was to be a 6:1 high-bypass-ratio engine to power the navy's carrier-launched S-3A antisubmarine aircraft. It used a core similar to the T64 engine and a fan and low-pressure turbine design scaled down from our GE1/6—the demonstrator engine that had already led to the TF39, the world's first high-bypass engine. After we won the competition with the GE12 to power the Sikorsky Black Hawk, we anticipated that the wealth would be shared, and Pratt & Whitney would power the S-3A. To our surprise, we won the competition with our TF34, an engine that was to have a long and prosperous production run. The TF34 would eventually power the A-10 Thunderbolt, affectionately known as the Warthog, which was still performing admirably in the Gulf Wars of 1991 and 2003. As the commercial version, the CF34 would power the Canadair Challenger in the 1980s and, more recently, the Canadair Regional Jet, which continues in spirited production to this day.

Although my job was called manager of production engineering, I had all of technical services working for me. I was working on the engineering aspects of these various projects and coordinating with manufacturing and visiting the airplane companies—and it was a fascinating, consuming job. I looked after all of these bright engineers working on new engines and derivatives of old ones while we were fixing field problems and helping to get the existing products manufactured. There were many people who were better engineers than I was—Art Adinolfi, who headed the design and mechanical teams for the GE12 and the TF34, and Art Adamson on the J85 and TF34, for example—but I was able to spot when something was wrong and pull a team together to fix it.

Not only was this job interesting, but it also brought me into a position of much greater visibility, although I have to admit that I traditionally found it hard to stay invisible for very long.

Gerhard Neumann

Gerhard Neumann was the leader of the Aircraft Engine Group at this time. He was a man of legendary charisma. Fleeing Hitler's Germany, he worked with Chennault's Flying Tigers in Burma and China. He sneaked into Japanese-controlled territory to patch together a downed Mitsubishi Zero so that it could be flown back for examination, then went back a second time after the pilot seriously damaged the plane attempting to land at a closer field. When the war ended, Gerhard, along with his wife and dog, traveled by jeep from Vietnam to Palestine. He liked African safaris, riding his motorcycle, and sailing. His management style was no less flamboyant. Behind his desk was a sign reading "Feel Insecure," and when he wanted to talk to the salaried work force, he would hold a tent meeting that looked for all the world like an old time, come-to-Jesus revival. A huge tent would be pitched in a parking lot, and the faithful would gather in the evening to hear the word and be saved.

At one of these meetings—I have long ago forgotten what was being talked about—I rose and spoke out in objection to what I thought was the wrong thing to do. A hush fell over the tent, and people looked at me as if I were a dead man. I have always believed that it is essential to stand up for what I believed was right. I expected and appreciated that in others, but I had no idea how Gerhard would respond. Nothing happened at the time, but many people thought I had dealt my career some serious harm.

Gerhard was always one to play psychological games with people, as if acting in slightly unexpected ways gave him an advantage in interactions. He was not a particularly large man, and my six-foot-four stature may have been imposing to him. His equalizer seemed to be his car. One day Gerhard invited me to lunch with him and his ever-present human resources man, Bob Miles—and he chose to drive. He had this tiny Renault that was frankly a cheap car. He and Miles rode in the front, and I crammed myself in the back seat. The lunch went well enough, even though I knew that I was what was being grilled. On the way back to the plant, I again squeezed into the back seat. When we arrived, I grabbed onto the door handle to help myself out, and, in utter embarrassment, I emerged with a broken door handle in my hand. I have no idea whether Gerhard's car was something of a surprise test he pulled on people, but I certainly could easily believe it.

Gerhard may have driven a cheap car, but he was not a penny-pincher in everything. He had a great sailboat that was his pride and the envy of many a GE employee, myself included. After I was well settled in my job as engineering manager, and earning enough to just afford it, Jill and I decided it was time to buy our own smaller boat. We went down to the marina, checkbook in hand, only to find Gerhard and Bob Miles walking out.

"What are you doing here?" asked Gerhard.

"We've come to buy a boat," Jill responded.

"A boat? What do you want a boat for?"

"Well, you've got to have a boat if you live in New England," I countered. "You've got a boat, and I would like one, too."

"I think you're making a big mistake," Gerhard said grinning. "I wouldn't do it, if I were you." I thought he was talking about the work and expense of owning a sailboat. What he probably meant, but did not say, was, "Where do you plan to sail that boat in Cincinnati?"

Summoned to Cincinnati

It was not too long after we had our 22-ft sailboat in the water—about four weeks later—that Fred Garry, the head of all production engineering, visited from Evendale and told me he wanted to take me out to dinner. Acting very secretive, he started with the gossip. He told me about a dispute that was going on between Jim Worsham, the head of military projects, and Fred MacFee, the head of engineering. He told me of plans to reorganize the business into functional organizations—engineering, manufacturing, engine projects. Then, like a bolt out of the blue, he said, "Brian, we want you to go back to Evendale to run Advanced Engineering."

I was stunned. "What would I want to do that for?" I replied. "I don't want to work in Advanced Engineering. I've already done that, and I love this job."

As we talked, it became clear that this was not exactly a request. I spent the rest of the meal thinking of ways to break the news to Jill. Our children were now 10, 9, and 7. We were all used to living in Boxford, about 20 miles north of the plant. We had even come to like the cold winters a bit and had learned to ski.

"Jill is not going to like this," I offered weakly.

"I'm sorry, Brian," Fred delivered the coup de grâce. "But we really do need you back there, and you have no choice."

I planned three or four approaches before I got home. I decided to go with the one in which I spend a couple hours of gently warming to the subject of moving, but I did not get a chance to use any of them. As I pulled into the driveway, Jill was standing at the doorstep. What have the kids done now, I wondered?

"Eric Doorly just called," Jill greeted me, without a hint of a smile. "He said he's taking your job. I understand we're going back to Cincinnati." So much for warming up.

While this process of announcing transfers seems a bit heartless in the description of it, this was not perceived as cruel or unusual. Just as it was in the military, when I was growing up in GE you were not asked if you wanted to go to the next assignment. When opportunities came, you took them. If you ducked what the bosses considered an opportunity, your chances of advancement lessened. And, usually, your family, whether they were pleased by the change or not, went along with what happened. I watched a few very competent people turn down transfers, and while they may have continued to grow, they never did as well as they should have. No matter how good they were, the bosses seemed unwilling to stick their necks out a second time.

Frankly, compared to the rest of GE, we were lucky in Aircraft Engines, in that most of us stayed in that business all of our careers. Gerhard tried to keep it that way. Jack Welch liked to see cross-fertilization across all of the businesses, and so he moved a lot of our people over the years, but these were mostly in manufacturing. I was actually quite fortunate to be transferred only twice. Bob Derochers, who was in finance, was transferred 17 times.

Lynn's contribution to the growth of GE Aircraft Engines

After I returned to Cincinnati, the team in Lynn went to work to demonstrate the GE12 and then to certify the T700 for the U.S. Army Blackhawk and the Apache helicopters, both of which proved to be excellent applications for the engine. At the same time the TF34 was being developed for the navy's S-3A antisubmarine aircraft. These military programs helped form the basis for two derivative commercial engines.

Early in 1980, Saab of Sweden started work on a 30-passenger twin-turboprop aircraft, in cooperation with Fairchild in the United States who were to build the wing. We competed with Garrett (Honeywell) for the engine for this application. The CT7, the turboshaft version of the T700, had been selected for the Bell 714 commercial helicopter and had also been subjected to many of the FAA tests during the T700 program. We proposed an engine with a gearbox mounted to the engine and separately supported by struts in the same way we had made the T64 turboprop work for the deHavilland Buffalo. After some very difficult negotia-

Earning My Stripes—Leading Engineers 35

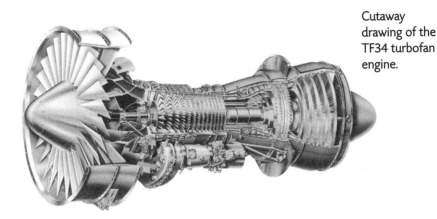

Cutaway drawing of the TF34 turbofan engine.

tions, we convinced a very tough Fairchild customer plus the Saab negotiators, who were always tough, that ours was the right engine.

Halfway through the program, Fairchild bailed out due to some internal difficulties and problems they were having with a trainer they were making for the U.S. Air Force. This put Saab in a bind as they already had some big commercial commitments that demanded some scrambling, but they handled everything very well. The Saab 340, with our CT7, turned out to be a very successful turboprop and sold throughout the world. It also brought us into contact with Dave Mueller of Comair, which at that time was a small, fledgling commuter airline operating out of Cincinnati. When Comair started I was on the board of the Cincinnati Airport, and we gave Comair the space that allowed them to put their two King Airs into service. After some initial problems, Comair really got into the swing of things, and with the Saab 340, they built a reputation for reliability. This relationship with Comair not only helped make the Saab 340 accepted around the world, but it also established the GE small engine business as being a reliable commercial engine supplier. This aircraft application also brought the Lynn engine people into the practice of supporting a big commercial fleet, which they had not done before but which they then did in exemplary fashion. Under the leadership of Dennis Williams, if ever the engine had problems, they turned the situation around very quickly without getting the rest of the organization in a turmoil. All of this helped create a greater expansion in the commuter market than we could ever have expected.

After the T34 went into service on the navy's S-3A, it was selected to be the engine on the A-10 attack aircraft, the famous Warthog of the Gulf Wars. During the initial service on the A-10, it became apparent that the hot section of the engine, i.e., combustor and turbine, needed considerable work. In actual missions, the A-10 had to go to takeoff power approximately seven times per flight compared with the S-3A, in which this occurred once. The fixes to offset the additional wear and tear on the engine took a lot of good engineering work and some special tests as well as some handholding of both the air force and the navy, but in the end, this process helped both the S-3A and the A-10 become very reliable airplanes.

Just after we had completed this work, Canadair of Canada was having some problems with the Lycoming engine on their Challenger business jet. As an additional plus, we could offer more thrust with our TF34. (The commercial version was later dubbed the CF34.) When the Challenger had been originally designed, there was a competition for the engine, and Canadair gave GE a tough time in the negotiations. Together with some people at FedEx—whom we had laid off and who bore us some ill will—they selected the Lycoming engine with its gear-driven fan. Now they were asking us to bid on the follow-on engine for a growth airplane. At the time this happened, we were deeply committed to spending a lot of money to develop the CF6 family, and so we were reluctant to bid. However, I felt that the growth airplane would be an excellent competitor to the Gulfstream, and so we made a proposal after reluctant agreement from GE headquarters in Fairfield.

We had just started the program to modify the TF34 into the CF34, changing the gearbox and improving the turbine, when Canadair came to us stating they had decided not to build the growth airplane. They did want the engine for the present aircraft, however, since sales had dried up because of problems they were having with the Lycoming engine.

I was really reluctant to build an engine for the present airplane because I felt sales were limited. As a result, we played hardball in negotiations and told them they would have to help us up front with the development costs, which we would pay them back proportionately with each engine sold. Although they really did not like this, they were stuck, and we cut a deal with them. We also said we would modify an aircraft at our test facility at Mohave, for which they also paid. This ultimately saved Canadair a lot of money, as our people did a great job on the program. (Frankly, as a critical element in developing all of our engines, the flight test people did a marvelous job on all of our flight test programs, and the CF34 program was no exception.)

Bob Kirk, who was running LTV at the time, built the nacelles for the engines. All this led to a very successful program for Canadair. It was a great thrill for me to fly in the first green Challenger being delivered to the completion center to have the interior fitted prior to delivery to GE. The airplane turned out to be a great workhorse, and that helped us develop a worldwide customer base. I found, as Neumann had before me, that the traveling required of a senior executive really demanded a business jet. (I am sure executives today find it even more difficult with all of the security delays on commercial airlines.) Because of this growing need, the Challenger turned out to be very successful and helped Canadair build an impressive reputation around the world.

Beyond the business jet market, it was apparent that the commuter business was in need of a turbofan jet commuter airplane as well. Canadair, together with our project managers, conducted a survey and found that there was a strong feeling by many airline customers that they would rather fly in a jet-powered airplane than a turboprop. We concluded that the Challenger could be modified to make a great commuter airplane with excellent passenger comfort compared to the turboprops to take advantage of the market demand. This took some commitment on our part since, if our engines went into this type of service, they would require significant modification to be able to fly reliably more than 2000 hours per year compared with approximately 700–800 hours that a business jet flew. We also would need to make other modifications to make the engine more adaptable for commercial operation. We decided we could do it, and after some protracted discussion

Earning My Stripes—Leading Engineers

with corporate headquarters in Fairfield, we committed. Little did we know that this decision would transform the commuter airline business and create a long-term growth opportunity for GE. Since that critical beginning step, the CF34 and its derivatives have built an outstanding reputation in the market. A growth version is now powering Challenger regional jets and the new Embraer aircraft and has just been selected for a new Chinese regional jet. These programs took some hard selling internally, but they worked out better than we ever expected.

Just after I left Lynn to return to Evendale, Eddie Woll, who was then vice president of Military Engines, felt the need for an engine to replace the J85 as Lynn's military fighter engine. He started his team on a demonstrator that would not be a small engine but would be what the aircraft engines community considered an Evendale-sized engine. This took a lot of selling inside GE, as issues of turf were being disturbed. The engine was intended to power a twin-engine fighter as part of an upcoming lightweight fighter competition. As it turned out, the winner of that competition was the F-16 with Pratt & Whitney's F100 engine, an engine that was already in the air force's inventory on the F-15. However, the U.S. Navy made a different choice. They selected the McDonnell Grumman F/A-18 Hornet, and GE's F404 engine (Eddie Woll's baby) was selected for this next-generation fighter. In the years to follow, the F404 proved to be a workhorse for the navy and many other customers around the world as well. The engine was subsequently grown in the late 1980s and early 1990s and became the F414 for the more advanced F/A-18 E/F. Besides powering the Hornet, the F404 was also selected by the Swedish Air Force and Saab to go on their Gripen fighter. Volvo is building the engines for this aircraft in Sweden. (It is interesting how recurring themes seem to have filled my life. One of my jobs as an apprentice was working with the Swedes on the Goblin and Ghost airplanes. By the end of my career, I could say I had had many great programs with them. The Swedes are good people to work with—after you have finished dealing with some of the toughest negotiators in the world.)

The chief engineer on the initial F404s was Frank Pickering, who was a key member of the Lynn engineering staff. Frank became vice president of Engineering for all of GE Aircraft Engines when I became president. Frank's leadership of the engineering team in the design of our engines during the 1980s through the early 1990s gave both Lynn and Evendale a great family of engines on which to build our future business.

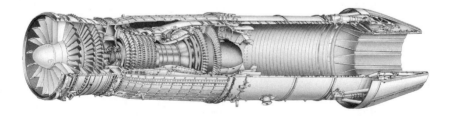

Cutaway drawing of the F404 turbofan engine.

When we restarted our commercial business in Evendale, Bob Kelly, the manager of finance in Lynn, was always joking with me as to how the commercial business was a loser and that we should get out of it. He appeared to be right during that startup period, and I am thankful that the military production programs at Lynn—T700, F404, J85, T64—helped us get through those critical startup years as we got our commercial business started. While they stung at the time, Bob's remarks made me very conscious of how much the growth of our total aircraft engine business was due to the contributions of the Lynn team.

My stay in Lynn was not only productively enjoyable, but it also provided my first managerial experience as well as my first taste of the commercial business. The members of the Lynn team were technically impressive, and they did outstanding work on some basic inventions. Throughout the rest of my career, they remained very special to me.

CHAPTER 4
Reentering the Large Commercial Engine Market

I learned in sports that, while everyone wants to win and hates to lose, if you do lose, you cannot dwell on the last loss if you want to win the next battle. As I returned to Evendale, GE had just suffered a big loss. The air force and the navy had asked GE and Pratt & Whitney to compete on engines to power their new F-15 Eagle and F-14 Tomcat. The contract would be a winner-take-all proposition. Pratt & Whitney won with their TF30 on the F-14 and their F100 on the F-15. Bob Hawkins was the chief engineer, and Jim Worsham was the project manager for our entry. Jim took the loss very hard, and he was riding everyone in the engineering community.

When I was appointed to head Advanced Engineering, I think it was partly to cool him down, and he took it as a great affront. He did not know who I was, nor did he care to, but he made it quite clear that he did not appreciate my presence. The first week that I was there was really tough. Jim was the kind of guy who would walk through walls to get the job done. In one instance, he burst in on Bob Hawkins working in his office and demanded that he come out and look at all of the vacant desks.

"No wonder we can't make it here," Worsham accosted him, pointing at the unpopulated office. "It's empty!"

"Jim," Hawkins tried to calm him. "It's 6:30 on Saturday. All the guys have gone home."

Gerhard loved—and rewarded—that get-it-done attitude. Worsham had put a huge amount of his own effort into the program and probably just needed to recoup a bit of energy. For the first few weeks, I just had to work with him and act as a shield between him and the engineering community. Everyone was amazed that I was bearding the lion that way, but they encouraged me when they saw that they were being spared the heat and could concentrate on doing their jobs. Worsham was not particularly pleased to see me in this role, and he told me so. Consequently, I was surprised when he offered to have his secretary work for me also. For about four months, I traveled back and forth between home in the Boston area and work at Evendale. When I finally moved my family and settled in, I found I had inherited Worsham's secretary, and he had gotten a new one. Apparently his former secretary had displeased him, and I was an easy solution to his problem. The secretary was a nice enough person, but I would have preferred to have had a choice.

Whatever friction his style may have generated, however, I have to admit that Jim's tenacity paid off many times in his career. When he was the leader of the GE1 program, he got the money to launch the GE1/6 program. Subsequently, his efforts kept McDonnell Douglas afloat with the sale of MD-80s to Japan Airlines (JAL).

In November of 1967, I had only been in the Advanced Engineering job a few months when Gerhard Neumann, the leader of the Aircraft Engine Group, called me to a meeting in his office. When I walked in, I saw that Ed Hood, who had been heading the supersonic transport engine project, was there, too.

"Brian," Gerhard began. "We're going to get back in the commercial engine business. Jim Krebs, Bruce Gordon, Bob Nietzel, and Bill Rodenbaugh have made some great progress modifying the TF39 and putting a low-noise fan on it as an engine for the L-1011 and DC-10. Ed's going to be my vice president for the commercial engine business, and we want you to head this engine project." This was to be the CF6, and was to be both a great experience for me and the right product at the right time for the business. As I look back, I now see that that short assignment in Advanced Engineering was to see if I could stand the heat and become an Evendale man again.

GE's large commercial engine business: a concise history

In the beginning of jet technology, there were only military engines. Then there were military engines with changes made to fit them to the needs of commercial aviation.

In 1950, a GE corporate committee headed by Jack Parker gazed into its crystal ball to study the future of the aircraft engine business. Parker had started his career as the head of the small engine business in Lynn and was in charge of all of GE's high technology businesses at the time. He and his committee recommended three courses of action: develop a new high-performance fighter engine, begin the further development of small gas turbine aircraft engines, and pursue the civil aviation market as well as the military.

Design work began on the J79 engine in 1952. This was to be an immensely successful high-performance military engine, and we would eventually produce more than 13,500 of them ourselves with another 3400 produced under license. At about the same time, Parker hired John Montgomery to head the Production Engine Department. Montgomery was a retired air force major general and had been running American Airlines' maintenance and engineering base in Tulsa, Oklahoma. Between 1952 and 1955, Parker, Montgomery, and Neil Burgess, the head of the J79 program, visited all of the major airlines in the United States and Europe to lay the groundwork for GE's entry into the commercial engine business.

This was still the era of piston engines, and the major players were Curtiss-Wright Aeronautical, Pratt & Whitney, and Rolls-Royce. Curtiss-Wright had some serious limitations. They had a reputation for being arrogantly unresponsive to field problems with their complex engines. They also had no commercial gas turbine engine in development. By comparison GE had impressive gas turbine technology, but it had no experience designing or supporting commercial engines. What little GE interaction there had been with airline customers in selling and supporting turbosuperchargers was a mixed blessing. The domestic airlines did want to see a viable competitor to Pratt & Whitney, however, Curtiss-Wright did not appear to be capable of rising to that challenge, leaving a potential opening for GE.

In 1952, British Overseas Airway Corporation (BOAC) began continental jet passenger service with the deHavilland Comet I powered by four Ghost engines.

Reentering the Large Commercial Engine Market 41

In 1954, after the accidents caused by airframe problems, the Comet was withdrawn from service and redesigned. The improved aircraft reentered the commercial market and transatlantic service began in 1958 with the Comet 4 and more substantial Rolls-Royce Avon engines. In the meantime, Boeing had developed the 707, and Douglas had the DC-8. Because GE did not have a commercial engine ready, both of these aircraft would be powered by Pratt & Whitney's JT3, the civil version of their J57, roughly comparable to GE's J79. Both the 707 and the DC-8 proved to be more efficient and have longer range than the Comet, and it soon faded from the market.

The DC-8 and 707 would work well for distant city pairs, but a number of airlines were looking for a jet aircraft that would be economically efficient for shorter routes. Convair had built successful piston-powered aircraft for this market decided to fill the niche. Convair had worked with GE on their supersonic B-58 bomber powered by four J79 engines, and they hoped the J79 could be converted to power the Convair 880, the medium-range commercial aircraft they envisioned. GE responded with the CJ805, a J79 minus its afterburner.

In 1956, Howard Hughes, the eccentric owner of TWA, launched the production of the new aircraft with an order for 30. Japan Airlines, Delta, and Cathay Pacific also ordered. Meanwhile, GE developed another version of the engine—the CJ805-23, which had an aft-mounted fan—for the somewhat larger Convair 990. American Airlines and Swissair launched this airplane, and SAS, Spantax, Varig, Garuda, and Thai Airlines placed orders. Unfortunately, a dispute arose between Hughes and Convair, and with the fighting and lawsuits, the 880 was delayed going into service. The other major carriers, instead of waiting, ordered 707s and DC-8s, first with JT3 turbojet engines, then with JT3Ds, Pratt & Whitney's front-mounted turbofan version.

In the end, the 880s and 990s proved too small for the market, and only about 100 Convair jetliners were delivered—about 400 installed engines. This exposed GE's reputation to performance, reliability, and other service-related issues, but did not provide enough income to make a profit. Convair lost a huge amount on the program and completely abandoned the commercial airplane market. GE also lost many millions, but continued to fully support these engines, even after the aircraft went from the major carriers into charter service.

Although the CJ805 was a derivative of the J79, it went into actual service within two short years after the J79 did on Lockheed's F-104 Starfighter. In their military role, these engines flew only 500 or 600 hours a year, and an engine removal every six months did not seem too bad. In the commercial market, however, engines would fly about 3000 hours a year, and no one wanted to remove an engine to maintain it. In other words, the military application gave little in the way of practical experience for the commercial market. As a result, we had to discover and fix the engine's shortcomings while it was in service with the airlines. The CJ805 entered service initially as something of a light, high-thrust, but potentially fragile, thoroughbred beside Pratt & Whitney's JT3 workhorse. We soon learned that the CJ805 was costly to maintain and not really rugged enough for airline service—with a plague of turbine bucket problems and excessive seal and bushing wear—but we stood behind it. It was vital for our long-term interests that we fix all of the CJ805's problems if we were to have any chance of credibility with the airlines. GE put extensive resources into the engine, which was doubly difficult since everyone knew it was a lost cause, and we eventually increased the durability considerably.

As a result, GE demonstrated to the worldwide commercial aviation community that it was prepared to back its products and its promises. We had discovered that our customers' appreciation for committed product support could give us strong competitive advantage. (We also ended up making the J79 a much better engine for the military.)

In November of 1967, when Neumann and Hood were talking with me about getting back into the commercial engine business, GE was still largely a producer of military engines. We did have a solidly profitable business-jet engine line, but this was really rather small in scope compared to the military programs.

The breakthrough technology of the TF39

In 1964, the U.S. Air Force had opened competition for engines to power an enormous new strategic cargo plane. Boeing, Douglas, and Lockheed each had a competing entry. The engines to power any version would have to be more powerful and more efficient than anything the world had seen to date. In addition, the Air Force was tired of being burned with cost overruns. The winner of this competition was to receive one lump sum to design, develop, produce, and support the engine for 10 years into the future. The intent of the Department of Defense was to shift the responsibility for bad decisions or poor productivity from the taxpayer to the manufacturer.

GE saw this as an opportunity to establish a benchmark for engine technology that might last into the next century. Taking a massive gamble, GE proposed a radically new concept, a high-bypass turbofan.

Early jet engines were turbojets. A rotating compressor compressed the outside air; a combustor injected fuel into the compressed air and ignited the mixture; and the hot exhaust, before it escaped to push the plane, used some of its energy to turn a turbine that kept the compressor going. A turbofan engine put another turbine in the hot gas flow. GE's first turbofans, the CF700 and the CJ805-23, were aft-fan designs. A second turbine was placed in the hot gas flow, and it turned a fan behind it at the very back of the engine that helped push the plane. Pratt & Whitney had built a front-fanned engine, their JT3D, but we had not. At the time, the CF700 and the CJ805-23 had the highest bypass ratios of any turbofan engines, 3:1, three times the volume of air drawn around the engine as goes through the hot core was to be mixed with fuel and burned. Turbofan technology is not limited to commercial engines. Some high-performance military engines, those that power aircraft that need quick bursts of power to get into or out of combat situations, have engines with an afterburner or augmentor. This is a long cylinder at the back of the engine with spray bars at the end closest to the engine. Raw fuel can be sprayed into the hot gas flow for a dramatic burst of power in flight or to help the plane take off. While the hot gases coming into the augmentor have plenty of heat to ignite the fuel being squirted into the augmentor, there is little oxygen left to burn. The purpose of the turbofan on these engines is to bring fresh air into the augmentor so that there is something to ignite the fuel. When the pilot is not using the afterburner, the bypass air does provide some additional thrust, but the bypass ratio on these high-performance engines is usually less than 1:1, that is, more air goes through the core than around it, and so the bypass air is less significant than it is on engines with a higher bypass ratio.

Reentering the Large Commercial Engine Market

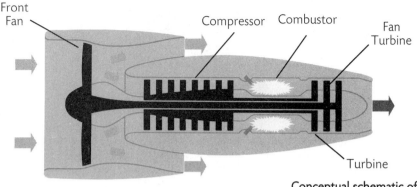

Conceptual schematic of a high-bypass turbofan.

What GE was proposing for the new military transport was an engine with a front-mounted fan that has a bypass ratio of 8:1. To make such an engine work, an enormous amount of energy had to be pulled out of the fuel and turned into a hot gas flow. This was necessary to turn the high-pressure turbine to power the compressor. Even more important, however, plenty of energy had to remain to turn the low-pressure turbine that powered the huge fan moving vast amounts of air through and around the core. To get that energy, the engine had to do a better job of compressing air than any previous engine—a pressure ratio of 25:1, much higher than anything that had been in service. Extracting so much energy from the fuel, the engine would have dramatically impressive fuel economy, twice as good as existing engines, but the gas flow coming out of the combustor would also be extremely hot. In fact, at just below 3000°F, the temperature of the gas flow would be significantly higher than the melting point of the blades in the turbine directly behind it. We would have to develop not only high-temperature alloys, but also exotic cooling techniques involving passages for cooling air inside these blades and minute holes over their surface to produce a cooling film of air over them just to keep this engine from consuming itself.

Turbine cooling has been a long evolutionary process that is still continuing, and is, in fact, a story in its own right. Our first experience with cooled blades had been for the J93 engine that was used on the B-70 Valkyrie. These blades had a set of radial holes going through them that were precision drilled. They worked well enough, but that engine did not really see much use. After that, we began using a lost-wax casting process that created serpentine passages for cooling air to go through the blades. We found that minute dust particles in the cooling air would accumulate in the passages, however. To let that dirt out, we began drilling little holes at the blade tip.

Then someone noted that we needed to keep the leading edge of the blades cool. We came up with the idea of film cooling—minute holes along the leading edge that tap into the internal cooling channels in the blade and allow a film of air to flow out to separate the blade that melts at about 1800°F from the 3000°F gases around it. Initially, we made these holes by electrolysis, essentially deplating very small diameter holes with electrolyte flowing through finely drawn glass tubes.

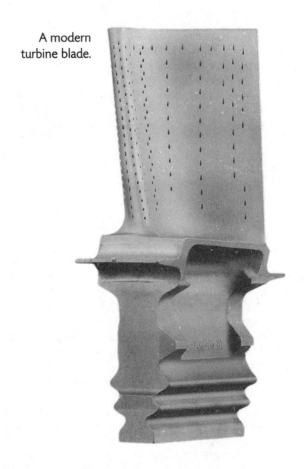

A modern turbine blade.

Later, we used lasers. While this sounds easy, a turbine blade is operating under extreme stresses, and we had to be sure we did not introduce uncontrolled stress concentrations at these tiny holes. Not only did we have to learn how to drill these less-than-a-millimeter holes in high-temperature alloys, but we also had to learn to chamfer the holes' edges as well.

After we had leading edge holes, someone pointed out that the leading edge is thick, and the thin trailing edge is more susceptible to heat damage. If we put holes in the leading edge, couldn't we also put them in the trailing edge? We ended up with turbine blades covered with holes.

As the materials improved, we also discovered that, when the molten alloy is poured to cast the blade, if it is poured and cooled in a certain way, the results are crystals that grow long and straight. These blades are much stronger than randomly crystallized castings. All this was a difficult job, but over time we did it.

Returning to our first high-bypass turbofan, the one-third scale GE1/6 demonstrator—the sixth engine variation using the GE1 core I had done compressor work on—ran and delivered all the performance we had promised, and a surprising specific fuel consumption of 0.336, less than half of that of the best engine

Reentering the Large Commercial Engine Market

then in service. Under the leadership of Don Berkey, the project manager of the TF39, who had Marty Hemsworth as his engineering manager, and with the tenacity of Jim Worsham, who was the chief of the GE1 building block engine and who headed the marketing effort, we drove to get this engine into production. With their good efforts we won a $459 million contract, the largest order GE had ever received for aircraft engines, for the full-scale engine that would be designated the TF39.

As a military contract, this was not the same business risk as it would have been if we were funding the project on our own in hopes of getting airlines to buy our engine. It was, however, a great technological risk. We had to do what we said we were going to do—something our competitors said was impossible—and do it in the relatively tight time frame of four years.

Lockheed won the airframe contract, and the new plane was designated the C-5 Galaxy. Boeing lost the competition for the military transport, but their work would not be in vain. The airframe they had designed could be modified to become an excellent large, fast airliner, carrying as many as 400 passengers on transatlantic routes at costs substantially lower than the 707s or DC-8s currently in service. Pan American saw the advantage such an airplane would give them and placed a launching order for 25 of the new Boeing 747s.

Boeing asked GE to propose a TF39-type engine for the 747, but this presented a dilemma. The TF39 design was optimized for the C-5. The thrust was too low to meet the Boeing specifications for the 747. Accepting Boeing's challenge would mean developing another entirely new high-bypass turbofan in parallel with the TF39. The younger managers were ready to accept this challenge, seeing it as a great opportunity to totally dominate the large aircraft engine market. Neumann was more cautious, however. Delivering what we had promised on the TF39 while keeping our other programs going would stretch our resources. Along with other projects, we were also working on a supersonic transport engine, a research and development program based on the engine for the Mach 3.0 B-70. Gerhard felt that another new engine program could cripple us if we ran into any difficulties. We

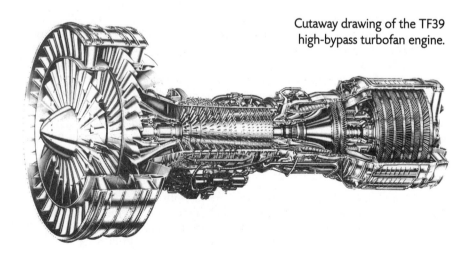

Cutaway drawing of the TF39 high-bypass turbofan engine.

might fail to deliver on both big high-bypass programs. Reluctantly, GE declined Boeing's request for an engine proposal on the 747.

Pratt & Whitney had also lost in the C-5 competition with their entry of a 4.5:1 bypass-ratio engine. In an apparent effort to block GE's threatened dominance in the big engine market, Pratt & Whitney offered to increase the thrust on their engine to meet Pan Am's requirements, no easy feat, and Boeing accepted their offer.

By the time I was meeting with Neumann and Hood in 1967, Gerhard was now ready to concede that we had the resources to develop a commercial engine based on the TF39. The meeting with him and Hood was to tell me that I would be the project manager for this engine, at that time a 36,000-lb thrust turbofan based on the TF39 core. Hood was to be my immediate boss as the vice president of all commercial engines.

The role of the project manager involves coordinating and directing everything about the engine—business strategy, design, development, testing, certification, marketing, manufacturing, service, and support—even though the people who provide these things may report to someone else in a complex business structure. Effective project management is an extreme example of teamwork in action. I wanted to play the game well—and win. So, just as I got on my bike as a boy to get a game together, I got busy pulling together a winning team—with a lot of help from Neumann, Hood, and the business staff—and sizing up the competition.

The commercial engine market of the mid-1960s

By the time we started the CF6 project, Pratt & Whitney was deeply entrenched on the 747, and so we had to look for other opportunities. Because of its dramatically lower cost per seat-mile, the 747 was about to revolutionize the economics of air travel. It was, however, much too big for the passenger load on many domestic routes in the United States. Major carriers were looking for a 250-passenger plane with the same seat-mile cost as the 747 to fly on those routes.

United, American, Eastern, and TWA had formed something of a consortium to push the creation of a new airplane to replace Boeing's 727, except with something more efficient and more reliable. Airbus was talking about a big twin-engined plane, but they did not have an engine for it. Douglas and Lockheed were also talking about big twins, but again, there was no engine to power them. The airlines would have preferred a two-engine airplane—the fewer the engines, the lower the cost and the easier the maintenance. At that time, however, neither we nor Rolls-Royce had the 50,000-lb thrust engine that would be required, and it seemed like an unreasonable stretch for us to promise one. Turning necessity into virtue, Jim Krebs and his team worked very hard to convince these four airlines to consider three engines instead of two on a new wide-bodied short-to-medium-range airplane.

Although the airlines were used to working with Pratt & Whitney, and the JT9D engine they were developing for the 747 might have had just enough thrust for a 250-passenger twin, Pratt & Whitney's engine was experiencing difficult problems. Not only had the 747 gotten heavier than planned, Pratt & Whitney was unable to raise the engine thrust to the level initially required by Pan Am. At full power, the engine casings were distorting from a perfect circle into an oval, allowing enough leakage around the various engine stages to reduce thrust and increase fuel consumption. Pan American refused to accept delivery of the aircraft until the engines met their specifications, and, at one point, Boeing had more than 30 undeliverable 747s parked on the ramp with huge concrete blocks hanging under their wings in place of missing engines.

Tri-jets: the Douglas DC-10 and the Lockheed L-1011

In light of the fact that no one really had an engine for a large twin-jet, GE's suggestion to the airlines that a three-engine airplane was not only feasible but also safer than one with two engines seemed reasonable. This set the stage for the design of the Douglas DC-10 and the Lockheed L-1011 trijets, airplanes that would be nearly twice as big as the 727. Knowing that the airlines were sometimes unhappy with both Pratt & Whitney's products and their product support, we went to American Airlines and United, the biggest North American carriers, and KLM in Europe and asked them to give us a chance on the DC-10. We had a good engine design, and we promised to give them first-class product support. Of course, we would have to go through a rigorous technical evaluation, but the airlines saw a lot of value in stirring up competition for engines and product support. That was all the encouragement we needed.

We were working closely with Douglas to optimize the performance of our CF6-6 for the DC-10 at about the same time that McDonnell Aircraft Company purchased them. We had worked very closely with McDonnell in the application of our J79 engine on their very successful F-4 Phantom, so our relationship with the airframe manufacturer was still somewhat cemented. The McDonnell family was quite conservative, however, and we were never quite sure if the McDonnells were going to support this airplane.

Mounting the third engine raised some questions. Boeing's 727, an earlier trijet, and Lockheed's L-1011 mounted the third engine in the end of the tail and used a complex S-duct to feed inlet air to it. While this arrangement was aerodynamically effective for the aircraft, the S-duct reduced the aerodynamic efficiency of the engine. Because our engine was a little longer than Rolls-Royce's entry, an aircraft with our engine in the tail would lose at least two potential rows of seats as well. McDonnell Douglas decided to mount the third engine in the vertical stabilizer on the tail of the aircraft. This cost a bit of additional structural weight, but overall it was a much better aerodynamic design.

Frank Kolk of American was one of the people concerned about the length of our engine. He was worried that, as an airplane with this long engine in the tail was backing out at LaGuardia, it would hit another airplane. (This never happened in service, by the way.) Because everybody seemed to be worried about it, we offered to warranty the problem. We also suggested to Kolk that he put flags on the tails of the airplanes to warn everybody. He did not appreciate the suggestion.

In the meantime, Rolls-Royce had proposed an engine, the RB.211-22, to compete with the CF6-6 on the Lockheed L-1011. Their three-spool design would be shorter than our engine, but it was also more complicated and an untried, from-scratch development project. As they have over the years, Rolls-Royce did a mag-

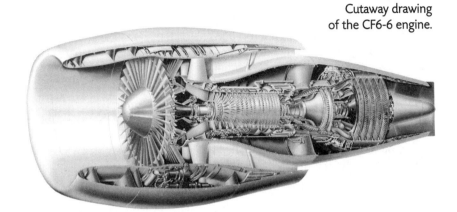

Cutaway drawing of the CF6-6 engine.

nificent job of pushing a little bit of technology and making it sound like a lot more than it really was. The British government would be putting up some development money, which helped spur airline confidence, and Rolls-Royce had faith in their carbon-fiber composite fan blade that we thought was a considerable technological risk. (It would run at a relatively high tip speed, and, frankly, the analytical tools necessary for properly laying the carbon fiber sheets as well as resin technology simply did not exist at the time.)

In what seemed either an ingenious marketing ploy or an act of desperation, Air Holdings Ltd., a new British aircraft broker, launched the Lockheed L-1011 with the RB.211-22 by placing an order for 50 aircraft. The British government offered export credit guarantees. (It turned out that Rolls-Royce actually was half-owner of Air Holdings Ltd.) Air Holdings immediately sold its 50 unbuilt airplanes to TWA and Eastern Airlines. The losses in GE's own backyard shocked us. Rolls-Royce had beaten us on the L-1011 without even a fight. We could not afford to lose on the DC-10.

Selling the CF6

Rolls-Royce had promised McDonnell Douglas an integrated engine-nacelle package. The nacelle, that streamlined enclosure for the engine, can be a complex, highly specialized piece of work. We really had not wanted to get involved in that part of the aerodynamic design, but we had to get McDonnell Douglas to accept our engine. Ted Stirgwolt, Ron Welch, and I got on a plane and went to Rohr Industries to convince them to design and build the nacelle. McDonnell Douglas would have a strong say in the design, but GE would be the system manager for the integrated nacelle. Getting Rohr to agree was relatively easy, but the people at McDonnell Douglas seemed to be dragging their feet, with no one willing to accept the responsibility to approve the package. After hashing out details until 10:30 one evening, I spent the rest of the night writing an agreement. In it, I described how we would work together, how we would make the integrated nacelle system work, how we would make it so that the engines could be overhauled on the wing, and then I signed it for GE. In the morning, I gave the agreement to the McDonnell Douglas representatives and said, "GE will stand behind this." This was subsequently known as the Chula Vista agreement and was one of the first *Working Together* programs in the industry. It ended up enabling us to build a much better nacelle and a more maintainable engine.

At the time, however, the McDonnell Douglas people hemmed and hawed and considered that they would have to show it to John Brizendine, McDonnell Douglas's leader of the DC-10 program. Ultimately, it was approved of course, but we had a great concern that delays were going to take us out of the running with the major domestic carriers. It certainly showed me that one thing we do well at GE is delegate responsibility.

With our nacelle package in hand, the GE team set about making impassioned sales pitches to Delta, American, and United on the merits of the CF6-6. First, we had to make them comfortable with the idea that GE could build and support commercial aircraft engines. Although we had a long history with military customers, our commercial experience was incredibly thin. We talked about the modular design of the CF6 and how the engine could be overhauled on the wing, and we explained the improvements we were making from the TF39 to commercialize it.

These improvements used to exasperate Gerhard. "I told you not to change anything!" he would blast at us, but the compressor rotor really did need a redesign. A military airplane, even a long-distance one like the C-5, might log 1000 hours a year, while most of the DC-10s would be operating well over 3000 hours. The rotor we proposed—actually, the way the rotor was held together—would be heavier, but it would last much longer and be more easily maintained. Fortunately, I had good support from all of the commercial engine people, and we came up with an all-new design we could be happy with. The military customers were eventually happy as well. When they adopted these changes, they ended up with a better engine, too.

Ed Hood led the way in marketing the engine. It took a tremendous effort on his part coordinating between the projects and Murray Ferguson and the financial people in New York to come up with a proposal that would fit the needs of both the airlines and McDonnell Douglas. As our reward, in 1968 American Airlines and United Airlines selected the DC-10 with our CF6-6 over the L-1011 with the Rolls-Royce engine. Had McDonnell Douglas not won this competition, it would have probably spelled the end of both the DC-10 and the CF6. Fortunately, McDonnell Douglas was a major producer of commercial aircraft at the time and probably seemed the safer choice. It turned out that it was also the wiser choice.

Ultimately, Rolls-Royce conceded that their costs on this engine had dramatically exceeded their target and the engine still was seriously overweight. Rolls-Royce went into receivership and the British government took the reins. The British government simply turned its back on the guaranties the Rolls-Royce had given to Lockheed and the launching airlines and told them that they had to pay whatever it cost to reduce the weight and bring up the performance. Lockheed, TWA, and Eastern were stuck—so they agreed. Lockheed had to seek federal loan guarantees just to survive.

At the congressional loan guarantee hearings, GE offered that it seemed inappropriate for the American government to guarantee loans to rescue a British engine while a perfectly acceptable American engine was waiting in wings. Needless to say, GE's position greatly angered Lockheed, TWA, and Eastern. The British government was furious with what it saw as GE's adversarial position, and a spirit of enmity lingered for years after that. The loan guaranties were granted, however, and Lockheed's aircraft went forward with Rolls-Royce engines. The L-1011 eventually became a technical success—though not a financial one. Thoroughly battered by the experience, Lockheed was never to make another commerical airliner.

Selling the DC-10 in Europe

Just as the domestic airlines in the United States had city-pairs that would never support the load factors to make Boeing's 747 practical, so too did the European airlines. Unlike the big American carriers, most of the European airlines were relatively small. To get the best deals on their aircraft purchases, a number of them would pool their resources into a buying syndicate. One syndicate was called KSSU—KLM, Swissair, Scandinavian Airlines, and UTA. These airlines had 747s, but they needed something else for their thinner routes. Although these routes had relatively fewer passengers, most of them were long routes with segments over ocean. A trijet made good sense, but the DC-10-10, the initial production aircraft, was a little too small and had a range that was a little too short.

To make the sale, Douglas proposed a bigger, heavier, longer-range aircraft, the DC-10-30, requiring three engines of 50,000 lb of thrust each. Ready or not, GE proposed just such an engine, the CF6-50.

We had just barely gotten the CF6-6 running. In October 1968, we fired up the first engine to test, and everyone was patting each other on the back. Then, after the third or fourth time we ran the engine, the chief test engineer forgot and left a stepladder standing in front of it. When we restarted the engine, it sucked the ladder in, totally destroying itself—hardly a good omen. We had started with an engine originally rated at 34,000 lb of thrust. As the DC-10-10 grew in development, the CF6-6 grew with it to 36,000 lb and then to 40,000 lb. Now, we were proposing to grow the thrust to 50,000 lb for the European carriers, with no real idea of how to do it.

Jack Parker was the vice chairman at GE by this time, but he still took a very active role in getting us back in the commercial engine business. Jack built relationships with all of the leaders of the KSSU and ATLAS groups, and so they were not surprised to see us competing. In addition, McDonnell Douglas had a reputation for building great airplanes, and they were well represented in negotiations by Dave Lewis and Jack McGowan. Rolls-Royce and Lockheed were also aggressively trying to win the order, and so there were many anxious people at the 1969 Paris Air Show where the decision was to be announced. The KSSU airlines were meeting at the UTA boardroom in Paris and continuing to negotiate with both contenders. Knut Hagrup and the SAS team were particularly tough. The Swedes seem to pursue negotiation as an art. I am not sure if they are as much concerned about the end result as whether they had what they feel to be a good negotiation. In any event, the questioning continued into the eleventh hour. Only when we saw the champagne being opened in the back of the room did we know we could relax and start giving noncommittal answers. We got the order, but now we had to design and develop an engine to perform as we had promised, a prospect of no great certainty.

Celebrating the sale of CF6-50 engines to the KSSU airlines with Gerhard Neumann (center) and Ed Hood (right).

When we first started to sell this engine, we considered that all we had to do to get from 40,000 to 50,000 lb of thrust was run the compressor at a slightly higher speed producing a slightly higher pressure ratio. After we did the calculations, we discovered this would raise the turbine inlet temperature from 2360°F to 2800°F. We had no experience at that time dealing with such high temperatures.

Luckily, Art Adamson had some thoughts. To reach the promised thrust level while keeping the turbine inlet temperature manageable, he suggested that we begin with the CF6-6 design, leave the fan diameter the same, but increase its speed, add three booster stages behind the fan, and take two stages off the back end of the compressor. This would deliver all of the gas flow needed to operate the turbines to produce the promised thrust while raising the cycle gas temperature just a little, still making this the hottest engine GE had ever tried to run, but not the white hot engine we had calculated. Art and his team made possible what seemed to be impossible and saved the business millions of dollars in the process. On the other hand, I still needed to sell this idea, and my first battle was with Gerhard. The product he wanted to be off-the-shelf was not off-the-shelf any more. This was now becoming an almost new engine. I was also to learn that you do not take a compressor you have been developing for years, supercharge it, knock off two stages of blades, and expect it to work the first time. When we started testing the supercharged compressor, we lost all of the now overstressed compressor blades. We had to modify them all, much to Gerhard's displeasure. We did end up with a good engine, however, one that would not have been possible without Art and his team's design.

Over the years, through a lot of personal attention from Jack Parker, we had cultivated some very trusting friendships in the KSSU airlines. Considering that our product may be in service with our customer for 30 years after the sale, taking years to establish relationships does not really seem out of proportion. Part of the KSSU trust of GE came from the fact that SAS and SwissAir operated Convair 990s with

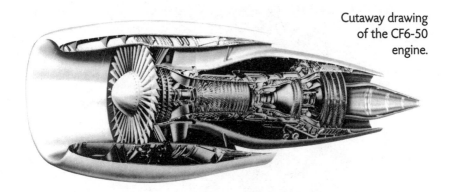

Cutaway drawing of the CF6-50 engine.

the CJ805-23 and got a sense of how we supported our products, even when we were not making money on them. In any event, KSSU trusted Douglas and GE to deliver. As a result, they bought the DC-10-30 with CF6-50s. This sale was shortly followed by another to the ATLAS syndicate—Air France (although Air France never used a DC-10), Alitalia, Lufthansa, Sabena, and Iberia. The two sales gave GE a big leg-up in Europe.

In the United States, we still had some convincing to do. Northwest Airlines needed a similar airplane, but they were openly disdainful of GE as an engine supplier. True to their feelings, they picked a version of Pratt & Whitney's JT9D for their DC-10-20s. It turned out to be a misstep for both of them. Very few airlines ordered the DC-10 with the JT9D-20. Pratt & Whitney had more than its share of difficulties with the engine, and the base was so low that there was no way to break even on the cost of fixing all the problems. GE had felt this kind of pain before and was happy to be an observer instead of a participant.

Customer support as a competitive advantage

GE was determined to leave the memory of the CJ805 program behind as we began the early stages of working on the CF6. In a move of dubious forethought, anyone who had anything to do with the CJ805 program was automatically excluded from working on the CF6 team even though they may have been the most qualified. As a result, when I looked at the impressive business plan that had been the basis of our economic decisions, I found something missing. The financial plan included nothing for product support, a disastrous oversight. After this near fiasco, we brought our experienced people into the CF6 program, regardless of their CJ805 connection.

One of the first additions was Dick Smith. Dick had been the product support manager for the CJ805 and, after that, for business-jet engines. He joined the team to set up a marketing organization. We wanted a systematic way to train our salespeople, to back them with financial analysis and sales engineering, to negotiate contracts and warranties, and to offer our customers sales financing. We had learned how to do these things on the CJ805 and the business jets. Now we needed to translate that to the CF6, find the right people for the jobs, and train them. We

Reentering the Large Commercial Engine Market

started with Cliff Whitbeck, Bob Boyne, and many others, including Harry Stonecipher.

Later, Walt VanDuyne joined the team. Walt had been the assistant director for sales and customer service for Wright Aeronautical. Wright had had a horrible reputation for customer service, largely because senior management had hamstrung their customer service organization. VanDuyne apparently did a lot of fighting for the customer, but never got his way at Wright. The experience had him loaded for bear. We gave him the responsibility for setting up our customer support function, and he and the rest of the team came back to me with requirements—so many people for repair engineering, so many for spare parts, so many for field service, so many for warranty administration. He laid out a training program that lasted as long as a year and a half for service engineers who had never been in the field before. He asked for the best people from our military field service program to flesh out the CF6's product support program. This was going to be hard-fought, expensive work—but I had had enough close connection with customers to know that treating them fairly and going beyond what we had promised was the only way to win the reputation that would make our commercial business a success. I approved the product support budget.

It seems strange that many businesses treat customer-contact jobs as relatively unimportant and insulate their leaders from the environment in which the products of their businesses are actually being used. This was certainly not true with us. We examined where Pratt & Whitney had not satisfied customers—in support, repair parts, manuals, timeliness of fixes—and we played to those weaknesses in our proposals, backing our promises with warrantees and cost guarantees. We stressed that we delivered what we said we would. Continual customer contact was the key to winning confidence. I used to tell the field service people, "You work for me. If you have a problem getting support, call me directly." And they did. Credibility was such an important issue for us that we went out of our way, sometimes completely beyond reasonable expectations, to keep customers satisfied.

The field engineers met every flight on the runway until the CF6 had proved itself in operation. Every morning at 7:30, I would call the CF6 project team meeting to order, and Walt VanDuyne, the Field Service manager, would report any problems that had come up since the previous morning. Then we would make assignments for corrective action on the spot.

Even manufacturing was a cooperative player. Normally, manufacturing is far more focused on getting completed products out the door than meeting spare parts needs, even if they are critical. Our operation was different. George Krall, the vice president and general manager of our manufacturing operation, considered himself to be part of the CF6 team and a party to our promise to keep our airline customers flying. There were conflicts now and then, of course, but he weighed the need to please our customers as heavily as we did.

Sometimes, I wondered if we went too far. Our product support people in the field not only felt responsible for ensuring customer satisfaction, but they also felt they had the power and the resources to get anything done. Whenever we found it necessary to raise our spare parts prices, the product support winced and cried and bled right along with the airlines. We often had a harder time convincing them than we did mollifying our real customers.

Airbus Industrie

Although European airlines were buying Boeing, Lockheed, and McDonnell Douglas airliners, European governments, especially the French and the Germans, were not particularly happy about the American monopoly in large commercial aircraft. With their encouragement, 1968 saw the advent of Airbus Industrie, a consortium of French, German, British, and Spanish aerospace companies pulled together to develop and produce European airliners. (Of course, this does not mean that Airbus content would invariably be 100% European. With engines and electronics, the American content of these European airplanes could be as much as 40%.) Their first product was the A300, a wide-bodied, two-engine aircraft—a little smaller than the L-1011 or the DC-10-10—for short- and medium-range city-pairs on the European continent. There was no comparable American aircraft.

The A300 was intended to be a philosophical statement of European technological prowess, and politics demanded that every possible element from design to hardware be European. The A300 would not only fill a European need; it was to be a mark of European pride to have a commercial airliner that could be exported to America and the rest of the world. A European Rolls-Royce engine was offered. We held little hope for getting an engine on the A300.

We had made a huge commitment to the commercial engine market with the CF6 for the DC-10, however; and European airlines had already ordered that application. A strong case could be made for engine commonality in their fleets making maintenance more economical. Even a small A300 program might give us some gain with relatively little incremental cost.

GE went to McDonnell Douglas with a suggestion. Cooperation with Airbus Industrie might go a long way toward making the governments of France, Germany, and Spain more receptive to the purchase of additional DC-10s. The A300 could use GE's CF6-50 engine with the McDonnell Douglas nacelle and mounting strut. We were willing to pay McDonnell Douglas $25 million to acquire the data and design rights for Airbus. (In retrospect, this was probably a big mistake for McDonnell Douglas in the long run, but it seemed like a good idea at the time.) We even persuaded Rohr Industries, the supplier of the nacelles for the CF6, to work with Airbus and to put a manufacturing shop to build engine nacelles in Toulouse, France, next to the Airbus assembly building.

None of this would have satisfied the need for a politically correct European airliner, however, if it did not also create work for European aircraft engine manufacturers. The major continental engine manufacturers were Snecma in France and Motoren-und Turbinen-Union (MTU) in Germany. We decided to enlist them as strategic allies. Our argument went something like this: If Rolls-Royce supplied the engine for the A300, it might parcel out some work to Snecma and MTU, but it would always be in charge. GE could offer them a more egalitarian arrangement.

GE, Snecma, and MTU had remarkably different work cultures and social strictures, however, and so comparing such things as plant productivity and other cost issues would be a huge problem. Because of lower volume and higher social costs, it would have been extremely difficult for either of these government-owned engine manufacturers to produce a CF6 part for less than GE could do it. While costs are what they are, however, the price is whatever the seller chooses to make it.

Reentering the Large Commercial Engine Market

With this in mind, Ed Hood proposed a novel concept to Snecma and MTU. They would make parts equivalent to 25% of the cost of the CF6-50 engine, not just the CF6-50s that ended up on A300s, but the CF6-50s on DC-10-30s and any future applications for the engine that European airlines would buy. For this, they would receive 25% of the selling price. How they ran their businesses would be up to them, and they were free to make as much profit as they could, or wanted, from their share of the sales revenue. While this had all of the appearances of a good offer, many French officials were deeply concerned about Snecma's ability to be competitive on a level global playing field. Snecma could end up losing money on every CF6-50 they participated in making, without any way to stop the flow of red ink.

As a further incentive, GE offered MTU the technology for drilling deep cooling holes in turbine blades, which they could use for their own programs as well. The only proviso was that MTU not join forces with any other engine manufacturer to compete against GE on the A300. They agreed.

Pratt & Whitney and Rolls-Royce were also contenders for the A300. Pratt & Whitney must have felt they had little chance or that this upstart airframe consortium was not worth much trouble. In addition, Pratt & Whitney was in great difficulty with their engine on the 747, and they had little credibility to compete for the Airbus order. They offered Snecma a standard subcontracting deal in which Snecma would supply parts to meet Pratt & Whitney's specifications, and Pratt & Whitney would mark them up and put them on the engine. They proposed their JT9D for the A300, but refused to make test engines available without cost.

Rolls-Royce placed its bet on the RB.211-22, their engine for the Lockheed L-1011. Their larger RB.207 existed only on paper. The Lockheed program was really consuming their attention, and they could see little advantage in dissipating their resources on an unproved European airframe syndicate. As Pratt & Whitney did, Rolls-Royce refused to provide free engines for flight testing.

When Airbus president Henri General Ziegler called Neumann about the free engines, Gerhard made a counterproposal. He offered Airbus eight engines for testing at no cost with the stipulation that Airbus pay for them after they had sold the test aircraft.

Although it seemed that we had covered all of our contingencies, we were still worried about one more competitive threat, that Rolls-Royce and Pratt & Whitney would form an alliance to offer an engine with Pratt & Whitney providing the engine and Rolls-Royce providing the European content. To forestall that, we offered Rolls-Royce a revenue-sharing deal similar to the one we offered Snecma and MTU. Rolls-Royce would have nothing to do with it. First of all, they wanted to develop their own engine to compete with the CF6-50, but the underlying reason was that they were convinced that our revenue-sharing proposal was a cunning plot to destroy Snecma and MTU—and they were not about to fall into it.

As the story played itself out, the CF6-50 was chosen to power the A300; Snecma and MTU made their share of the hardware in one of the most profitable programs either of them had ever enjoyed; and GE became the dominant supplier of engines for Airbus Industrie aircraft. It worked out well for Airbus, too. Their mid-sized twin made the trijets uneconomical. Boeing decided to follow Airbus's lead with their own mid-sized twin, the 767, but Airbus was clearly established in the world market. Within 15 years, they would surpass McDonnell Douglas as the second largest airframe manufacturer and now have virtual parity with Boeing.

Engines for Boeing

When Boeing had asked us to develop an engine for their 747, we declined. Now that we had the resources and the engine to do the job in our CF6-50, Pratt & Whitney had captured the initial market. Unfortunately for them, the problems they were having with their engine were not winning them many friends at the airlines. Potential new customers for the 747 were asking Boeing if they could have the airplane with GE or Rolls-Royce engines. Boeing's response was that, while that was possible, it would mean recertifying the airplane with the new engine combination—an extremely costly and time-consuming proposition. No one wanted to bear that cost for a small order, and there were no large-volume orders on the horizon that might be able to make that cost acceptable.

T. Wilson, the new chairman of Boeing, and Jack Parker from GE met and talked about it. They decided that each company would put up half the money toward retrofitting a 747 with GE engines and certifying the airplane. The 50,000 lb of thrust of GE's CF6-50 would be 8000–10,000 lb more than Pratt & Whitney's engine, significantly increasing the 747's capacity. So we put four engines on the airplane, and it really performed well. We expected that we might sell about 25 aircraft in this configuration, but we still did not have a customer.

Fortunately for us, the U.S. government wanted to replace *Air Force One*, the president's aging 707 executive jet, with a new 747-type aircraft. At the same time, the air force wanted to procure several similar aircraft as airborne command posts. This new application would be designated the E-4. Compared with a standard 747, the E-4 required an enormous amount of electrical power for all of its communication gear. Because we had been so successful working with the air force on the C-5 Galaxy powered by our high-bypass TF39 turbofans, they welcomed our proposal to provide CF6-50s with special gearboxes to deliver that extra electrical power to power the E-4. The test protocols of the E-4 program would provide a lot of good reliability data, and so this program could be something of a "free ride" into the 747 market.

Wilson was not entirely convinced that we could handle this project. The E-4 would be a high-profile, showcase application, and the last thing he wanted was for Boeing to be embarrassed again by an engine supplier. He called Jack Parker and Gerhard Neumann to test GE's commitment to the reliability of the CF6-50.

We had been providing the engines for the VIP helicopter squadron that ferried the president and his cabinet for some time. We had even developed a special product support system for these engines. Gerhard and Jack promised Wilson that we would provide the same level of support to the E-4, and Wilson agreed. We were very lucky that Wilson selected Roy Anderson to head up the installation of the CF6 on the 747. Roy did a terrific job, and the shared cost between GE and Boeing really paid off for both of us.

Boeing offered the 747 with GE engines to their commercial customers, and an opportunity came quickly. Lufthansa wanted to purchase new 747s to replace their older ones. They already had DC-10-30s and A300s powered by our CF6-50 engines. It made economic sense to have the same engines on their 747s. Not only would they save money from having a common parts and spares inventory, but also their pilots and mechanics would benefit from familiarity with the same engine throughout their fleet. As a result of this compelling logic—and our past experience working together—Lufthansa bought the first commercial 747s powered by GE engines. At the same time we obtained an order from KLM. Soon Air France and the other KSSU and ATLAS airlines followed KLM's and Lufthansa's lead, as did 14 smaller airlines that look to the bigger carriers for technological direction.

While we entered the 747 market with a fury, we certainly did not eliminate our competition. Northwest and Japan Airlines continued to be loyal all-Pratt & Whitney purchasers. Other users, such as Singapore Airlines, took advantage of the fierce competitive pricing that Pratt & Whitney was using in an attempt to turn tide to garner some bargains. Boeing certified Rolls-Royce's RB.211-524 on the 747, and British Airways, Air New Zealand, Air India, and Cathay Pacific ordered 747s with Rolls-Royce engines.

Boeing found that they could expand their potential market by being able to offer an airplane with a choice of three different engine suppliers. This fact was not lost on Airbus Industrie. They soon started offering engine choices on their airliners. Even though the majority of Airbus aircraft continued to be ordered with GE engines, our informal exclusive arrangement was over.

While all of this market expansion was taking place, my CF6 team was producing engines that met the promises we committed to. For me personally, this meant extensive travel to help sell engines and keep customers happy. On the home front, however, I had outstanding support from Harry Stonecipher on the CF6-50 and Wally Dodge on the CF6-6. Most importantly, our teams were building a reputation for good product support, which made meeting with the airlines much easier.

The CF6-80 series

In the late 1970s, Airbus was designing the A310 and Boeing was working on the B767. Both of these aircraft were to be slightly smaller than the A300, so their power requirements were initially below 50,000 lbs of thrust. In the meantime we had de-rated the CF6-50 to 45,000 lbs of thrust for use on a special short-range version of the B747 for ANA. Because we had this engine in hand, we proposed it to Airbus and Boeing. This had the additional advantage of providing commonality with our engines around the world. The engine proposal was well received by Airbus as well as by Lufthansa, which was to be one of the launch customers.

Joe Sutter, the chief engineer of Boeing's 767, wanted a new engine, however. Bob Hawkins and I went to Seattle and saw Joe to be sure of what he wanted. While we were there, we came up with the idea of taking the turbine mid-frame out of the CF6-50 and redesigning the low-pressure turbine to reduce the length, weight, and complexity of the engine. We also added some better performance features. This engine, designated the CF6-80A, was to be the first engine of the CF6-80 series, which was to be the best of the CF6 family. It went into service on both the B767 and A310 in 1983. Hawkins and I returned to Cincinnati to get agreement from Fred MacFee and Lou Tomasetti, who were running commercial projects at the time. Together they sold the idea to Delta Airlines, which was one of the launch customers of the 767.

The reliability of the CF6-80A was not as good as we had wanted, but it was a good start to this series. Neither the initial 767s nor the A310s sold that well, and both Boeing and Airbus started to market newer versions. For their A300-600, we proposed an engine called the CF6-80C to Airbus. This was to be an up-rated CF6-80A with some additional boost and a slightly bigger fan. In the meantime Pratt & Whitney had offered the PW4000, which was to be a much better engine than its JT9 that was currently being used in the various aircraft. Rolls-Royce offered the RB211-Trent, which was already in service on B747s.

As we got into the development of the CF6-80C, it became apparent that it was not good enough to take on the PW4000, and it needed to be improved. In response, we increased the fan diameter and improved all of the various components of the engine. These changes meant that we had to delay the certification of the engine, which also delayed the A300-600 certification. Airbus was furious but also saw the potential of having a much better airplane—an airplane that continues to sell well in the 2000s.

In the meantime, Boeing told us they wanted no part of this change as they didn't want to see the A300-600 get better. At the time, they had no intention of stretching the B767. That attitude changed very quickly, however, when they decided to stretch the 747 first to the 747-300 and eventually to the 747-400—which turned out to be the best of the 747 product line. They also used the more powerful CF6-80C on the 767 long-range aircraft. It turned out to be a good fit as well for the MD-11—a stretched MD-10-30 with winglets and a higher gross rate. The CF6-80C is still in production today, with more than 3300 sold. It is also one of the most reliable engines in its service category.

Our biggest single sale with this engine was at JAL, where we had not sold any engines since our restart in the commercial business. We were helped by the impressive success the engine had at ANA, the other major Japanese carrier. Our team, headed by Ed Bavaria and Sandy McCord together with our Japanese office representative, did a tremendous job in the campaign that lasted over several months. We also had great help from our service people who played a major role in convincing JAL's service and engineering people that we were there to serve them.

We also produced the CF6-80E for the A330. This engine was rated at 68,000 lbs of thrust.

The CF6 engine family, which started in 1967, is still in production today. This engine family was the bedrock from which all of our commercial engines gained a reputation for performance and reliability around the world. It put us in the big league. Six thousand four hundred were sold, but the biggest seller was the CF6-80C. It has a tremendous reliability record, helped by one of the best compressor and turbine combinations in the world.

CHAPTER 5
Love and Hate—
Working with the Military

GE entered the aviation business while World War I raged in Europe, when Sanford Moss designed an air-cooled turbo-supercharger to enable military piston aircraft to fly higher. The design was tested on Pikes Peak to demonstrate its capability and durability. Based on the tests, the military gave GE a big contract for turbo-superchargers, but the war ended before production began, and the contract was canceled. (GE's turbo-superchargers did play a major role in World War II, however.)

GE's aircraft engine business started with the military as well. General Electric built America's first jet engine for the Army Air Corps in 1942—the centrifugal-flow I-A that used Sir Frank Whittle's British design. GE was selected to do this because of all of their experience with turbo-superchargers. The production run was 15 engines.

In 1943, GE designed, developed and built the 4000-lb thrust centrifugal-flow J33 that powered Lockheed's P-80 Lightning. (At the war's end, the government turned production of the J33 over to the Allison Division of General Motors.) In the same year, GE also designed and developed the 4000-lb axial-flow J35 for the Republic P-84 and Northrop's flying wing. (Citing GE's limited production capacity, the government awarded prime production responsibility to Chevrolet and, when Chevrolet started building cars again, to Allison.)

In 1946, the Supercharger Engineering Department at Lynn redesigned the J35 to deliver 5000 lb of thrust. This new engine, the J47, was installed in the F-86 Sabrejet, the B-45 bomber, the B-47, and later the B-36. Thousands of these engines were needed for the Korean War that broke out in 1950. To fill the need, GE not only built J47s in Lynn, they also opened the Lockland Plant in a former Wright Aeronautical site near Cincinnati. (This is now the Evendale facility, headquarters of GE Aircraft Engines.) In 1949, GE grew the J47 into the J73, an engine delivering 9000 lb of thrust, for the F-86H. The J35-J47-J73 families of engines were exclusively military products.

The year 1954 saw the introduction of an entirely new engine by GE, the J79 with 10,000 lb of dry thrust, that is, thrust generated without squirting raw fuel into the hot gas flow in an afterburner, and 14,500 lb with such an afterburner. This engine powered Lockheed's F-104 Starfighter, the B-58 Mach 2 bomber, and later McDonnell Douglas' F-4 Phantom. Seeing an opportunity to enter the commercial engine business with the J79 core, as previously mentioned GE developed the CJ805-3, a turbojet for the Convair 880. This was the mid-1950s. I had gotten my degree from Durham and was busily working on rocket motors at deHavilland and considering a sojourn in America at the time.

When I had started as an apprentice for deHavilland, virtually all of our engines were military engines. The contracts for those engines seemed to be parceled out to the various manufacturers as they might have been for any commodity such as shoes or tents or gasoline that the military purchased. At my level, there was no contact with the military customer, and I never saw anyone from the army or the Royal Air Force around the plant. How different it was to be in America.

When I arrived at GE in 1957, it became clear that the government was the toughest customer imaginable. Competition for military orders was intense and elaborate, and specifications were strict and inflexible. While a competition for a commercial engine order today might last three months, a military competition sometime lasted three years. The intensity of competition among engine producers often generated many marvelous breakthroughs, just as the customer hoped it would, but it was enormously frustrating when we awoke to the fact that our government customers legally owned every idea, every plan, every drawing, every fruit of our effort on their behalf.

When your ideas are not your own

There were a few instances in which this really hit home. In the mid-1960s, we developed our concept for the TF39 as a novel very high-bypass engine to power the C-5 Galaxy strategic transport. This was a risky design in that it broke new ground with an 8:1 bypass ratio. Bypass ratios even close to this for production engines had never been tried. Driving the huge fan to generate all that bypass air on this engine demanded turbine inlet temperatures that had not even been attempted. As noted earlier, we invented material and manufacturing technology to produce turbine blades cooled by a thin film of air that passed through channels inside the blades and emerged from minute holes on the blades' surface. Others in the industry thought it would never work—Pratt & Whitney had proposed a much more conventional engine for the C-5, and it was in their interest to point out the risk in our plans—but we developed the idea and tested it and proved the concept. Our military customer then took our concept and our data, made them available to the industry, and asked for competitive bids. When you work for the government, you do not own your ideas.

Sometimes, this customer will make that point forcefully even if you win the competition and your engine is in production. We were producing F404 engines for the U.S. Navy's F/A-18 Hornet. In the late 1980s, the navy figured that competition might lower their costs, and Pratt & Whitney told them they could build F404s for less than we could. The navy gave Pratt & Whitney the new order, along with a $400 million subsidy to set them up in production. Pratt & Whitney never came close to meeting our price much less beating it, and in the end the whole production came back to us. In the meantime, taxpayers had spent $400 million with nothing to show for it. At that time, critics in government, such as Sen. William Proxmire, were busy looking at the cost of toilet seats and coffee makers and not the money wasted on big political items!

Up the bureaucracy

Like any other customer, military customers certainly want performance, reliability, and maintainability in the products they buy, but these all seem to take a

Love and Hate—Working with the Military

back seat to cost in the public forum. Government projects have a reputation for being preferentially mismanaged just as defense contractors are considered to be perniciously greedy and amoral in their pursuit of the taxpayer's pocketbook. To overcome this impression, be it real or imagined, the government and its agencies have rigid rules to manage bid proposals, development, and production. The various levels of checks and balances are there to ensure that there is no such thing as a rigged or preferential bid, that the product as defined in development does exactly what it was promised to do, and that the product produced on the assembly floor is precisely the same product approved after development.

To ensure that this happens, there are volumes of rules and seemingly endless approval cycles for any changes. The penalties for violating the rules or procedures can be severe, even if the result is a better product or a saving to the taxpayer. As a result, GE had to install its own expensive bureaucracy to act as an interface with the government one. This internal bureaucracy developed and expanded over the years until, by the time I joined GE, the aircraft engine business was fairly fat, with many levels of management. I remember how shocked I was coming from the lean deHavilland operation to find the many managers whose jobs had little to do with the actual design and manufacture of jet engines.

Because of the continual specter of excessive profits going to unscrupulous contractors, the government has tried many ways to limit what a manufacturer could make. Initially, that was the cost-plus contract. The manufacturer would do a rigorous cost accounting to be approved by the government and a percentage of that cost would be added as profit. Clearly, under this system, there was no incentive for the manufacturer to reduce costs. That responsibility fell to the government, which could challenge what it felt were inappropriate or excessive costs.

To make the acquisition process more cost effective, the government moved to fixed-cost contracts. Fixed-cost contracts were based on the same mutual understanding of costs plus what the government considered a reasonable profit, but the dollar amounts were committed to at the beginning of the contract, rather than subject to an analysis of actual costs.

The good news for the U.S. Treasury was that the amount of the contract was locked inflexibly at a certain dollar amount. The bad news for the military customer was that it was almost impossible to make changes, even changes that would dramatically improve the product, if that meant the cost would go up. As you develop something as complex as an aircraft engine, invariably the world continues to move, and your requirements change, but re-negotiating a fixed-price contract is an ungodly mess, and so both parties just live with it as-is if they can.

Whenever we would win a fixed cost military contract, we would work ferociously to reduce our costs of production. On one hand, our cost-reduction strategies allowed us to make a larger profit, and by driving down our cost base it made us tougher competitors for the future. On the other hand, the military customer, by regulation, knew as much about our cost system as we did. The next time orders came up for renewal—and usually that was annually—we would have to negotiate a new price based on our actual costs so that the government could reduce our profit to what they thought was acceptable. Then the cycle of cost reduction and subsequent negotiation would begin again. After a few years our people were sure we were trying to squeeze lifeblood out of a pretty limp turnip, but we got exceptionally good at reducing costs.

I know that many people think government contracts are like a license to print money, but they are not exactly the sweet deal the public at large imagines they are. To be fair, I should point out that, if you do a good job, you are almost guaranteed a profit. If you make mistakes that cost you money, no customer—not even the government—will be sympathetic. If you run into totally unexpected technical problems, however, the military customer will usually help fund the fix, as opposed to the commercial side of the business where, if you have a problem, it's yours.

Myopic focus on cost

Unfortunately, the *cost* tail often wags the *total value* dog in government accounting. One of the worst examples from my perspective was on the C-5A, the first production run of Galaxys. Cost overruns caused by "totally unexpected technical problems" on past projects had been such a public relations problem that Secretary of Defense Robert McNamara had made an issue of the C-5 being a total-cost program—a fixed lump of money for the program from beginning to end.

This airplane was a brand new concept, however. The C-5 was huge, bigger than any airplane built at that time. As Lockheed got into development of the airframe, it was obvious that the wing had a problem. It needed to be strengthened, but that would add weight. We suggested that they beef up the wing, and for a minimal development cost, I think about $50 million, we would increase the thrust of the engines to compensate for the increased weight.

The government's answer was a resounding no! We were to keep the thrust as planned and Lockheed would make the wing design work within the allotted weight, with no increase in cost. Ten years later, the government accepted the fact that the air force would have to re-wing these planes to make the wings stronger. The cost was $5 billion.

With commercial projects, whenever we would see a way to make an improvement that would reduce the customer's total cost of operation, we would simply include those innovations. Often, these better ways of doing things saved everybody money, the customer and us. With the military, the decision to introduce a change was always hampered by the inertia of an enormous bureaucracy. Changes that required a small investment today for enormous payoffs tomorrow were often dismissed.

The T58 presents a few good examples of this. This is a helicopter engine we made for the navy. We would see problems that the operators or the mechanics were having in the field and create fixes to overcome them. To install the fixes would cost some money then, but not installing them would cost even more money in the long run. The year-to-year budget would not have anything in it for fixes, however; and so we would sit in program reviews with the navy and talk about the problems. We would have a fix; they would know we had a fix; but both of us knew they were not going to fix it. From an engineer's perspective, there is nothing more frustrating than solving a problem and having that solution ignored, especially if serious consequences could result from the problem. At one point, we had 58 proposed fixes to resolve in-service problems, fixes we thought were necessary but could not introduce. When the contract came up for renewal, we included those changes of course, but this compromised interchangeability in the field.

The public typically sees defense contractors as an irresponsible lot, gorging themselves at public expense with one cost overrun after another. I am sure there are contractors who are either unscrupulous or criminally careless, who consistently bid a job for less than it will cost, expecting that the government will bail them out to protect the program. I cannot remember us ever going back to a government customer and asking for more money because either we had underestimated costs or financial conditions changed. We would simply absorb any underestimation and make less or no profit—which did not make the GE hierarchy happy—but in the end we usually made up for it.

Playing both ends against the middle

We always try to protect ourselves from financial mistakes. We would love to put in escalation clauses in fixed multiyear contracts to cover actual changes in the inflation rate, but if the customer demands a firm fixed price for a number of years, we estimate what inflation will be and plug that into our quote as part of our risk. On one program, the U.S. Air Force wanted just such a firm fixed cost. We made our estimates and did some negotiation, and ultimately the air force and GE agreed on a price, but we were still very concerned as inflation was running rampant at the time. As luck would have it, the rate of inflation was less than our estimates. The air force came back to us and wanted the price reduced accordingly. We said, "Wait a minute. If inflation had been higher, the additional money would have come out of our pocket, not the taxpayer's. The fixed price was the whole point of having a contract."

They did not see it that way. It took John Lehman, who was Secretary of the Navy at the time, to point out to the government procurement people that either they did or did not want a fixed-price contract. They could not have it both ways, fixed if that benefited them or fluctuating if that benefited them. We did not give the money back as such, but a customer is a customer, and we did some creative renegotiation on the whole thing to make both sides happy.

Working for a government customer is inherently less efficient than building commercial engines. For one thing, there were all of the people in your plant representing the taxpayers' interests. They would walk around the plant, checking to see if the people we said were working on the program were actually doing that. They would inspect the parts we said were not fit for service to see if we were actually throwing away good material. They would look over our shoulders and scrutinize the results to see that each engine passed all of the tests. They would examine the parts list in the shipping documents to verify that the part numbers we agreed to ship were in the engine. They would check everything. When you have 300–400 military people in your plant asking questions all day long, you have to add about 1000 people whose jobs are to find answers to those questions. On the commercial side, we did not have a bunch of people always answering questions like that. (The Federal Aviation Administration [FAA] had only one representative in the plant.)

If you come up with a change that would make a military engine better or safer, there is an enormous bureaucratic process to wade through to get the change approved. On the commercial side, we would just make the change on our own—

although we may have to subsidize installing it in engines in the field—and, of course, we would have to get FAA approval and do the appropriate testing.

When I came to GE in 1957, the CJ805, GE's initial entry in the large commercial engine business, was under development. Essentially, everything else was military. In the military programs, there were layers of people checking each other's work. The obvious reason for all of these checks and balances was that nobody wanted to make a mistake, but as a result of this shared responsibility, few individuals stepped forward to take personal charge. The military customer was the final approving authority on anything we did, and so we had gotten used to deferring decisions and trusting the workings of red tape. Overcoming this aspect of the business culture was one of the biggest tasks I saw before us when we reentered the large commercial engine business in 1967 with the CF6. We had a tough job getting people to think, "We are the ones responsible; we don't have to get somebody else's approval," and then being ready to stand up to that responsibility.

From military to commercial

Of course, military engines were our springboard into commercial engine production. The CJ805 engines for the Convair 880 and 990 came from the J79 core engine; the CJ610 and the CF700 for executive and business jets came from the J85; the CT58 for commercial helicopters came from the military T58; the CF34 on the Canadair Challenger and Regional Jet came from the TF34; the CF6 family began with the core technology of the TF39; the CFM56 family sprung from the core of the F101. Only the GE90, our engine for the Boeing 777, was developed specifically as a commercial engine. In all cases, however, fixes and improvements made for the commercial engines were fed back into the military engines, and as a result the TF39, TF34, T58, J79, J85, F101, F110, and F404 became much better engines.

Some would look at this and suggest that we operate in a heavily subsidized industry. That is a surface appearance only. For one thing, the core engine—the gas generator consisting of the high-pressure compressor, the combustor, and the high-pressure turbine—is only half of a turbofan engine. The fan/low-pressure compressor and low-pressure turbine for a commercial engine still need to be designed and developed. The requirements of the low-pressure system often demand that we redesign some of the high-pressure system.

Commercial engines must run quietly, and so there is sound dampening or acoustical design changes to be added. The protection of passengers is critical to airlines, so that containment structures must be added to protect them in the unlikely event that the rotating parts of the engine should break. Then there is the cost of actually certifying the engine on the commercial aircraft application; this is a long and expensive process. Finally, we pay the government a royalty for the use of technology that we developed while working on their engine project. So, the transition from a military to a commercial application is hardly a free ride. It does help a lot with developing the basic technology, though.

Transferring military technology to commercial applications also has its counterpart. Commercial engines are operated almost continually to get the greatest return from a large investment. Military engines are operated for training and for

Love and Hate—Working with the Military

operational necessity. In the absence of war, the cost of fuel and maintenance keeps military aircraft on the ground much more than commercial aircraft.

The result of this is that we get considerably more knowledge from experience with commercial engines. We may change a part design to improve performance, increase life, or reduce cost. When it is time to replace parts in an overhaul, the military often benefits from what we learned in the commercial marketplace. We essentially continue to develop their product even though they have stopped paying us for that kind of work.

The differences between the military and the commercial market often make them seem like night and day. Military contracts offer low, but somewhat assured, profits that you see relatively immediately. From an engineer's standpoint, they can be both exciting and frustrating, exciting because you are usually dealing with cutting-edge technology, frustrating because your hands are often tied when you come up with a better way of doing things at the wrong time. From a business operation standpoint, it is often difficult to determine just who is the customer. Sometimes, the awarding of government contracts seems to be strongly influenced by congressional politics, so a congressman or a congressional staff member may be your first customer. If you get the contract, the auditors, overseers, and bean counters take over, and your customer is a legalistic financial and quality system. Finally, as your engines get to the field, your customers are the people who maintain and fly them. With military engines, it is much more likely that each of these customers will have different needs and that we will be placed in the position of having to satisfy them sequentially, rather than all at once from the beginning.

In the commercial market, there can be high profits, but they may take as long as 20 years to develop. There are no assurances, and you have to have deep pockets to weather any setbacks. The cost of developing an engine is entirely your own, and that cost can easily reach $1.5 billion over a four- or five-year development cycle for a totally new engine. Competition determines the price, and to enter a market you may have to launch your engines at a price below your estimated production cost and hope that you can reduce costs enough to break even or make a small profit. Sometimes the early market never improves, and competitors continue to sell at little or no margin, or sometimes at a loss, to gain market share.

Market share is very important in the commercial business, because spare parts sales are so lucrative. While the margin for engines might be small or nonexistent, the margin for replacement parts is high. Some may see this as gouging a captive customer—certainly many of the airlines see it that way—but the airlines have no difficulty negotiating give-away prices from us for new engines. Finally, the needs and wants of our commercial customers seem to have a greater consistency throughout the phases of design, development, manufacturing, and use. That consistency is built on the profit motive. As a result, if we show them how they can spend money to make more money, they usually listen. It would be better for the future of the engine business, of course, if the customers would pay more up front for engines in order to have reduced spares prices. It seems quite an imposition to ask the engine manufacturer to carry a program for 15–20 years before it breaks even.

A changing attitude

Our military customers have seen what the commercial marketplace has done to engine prices, and they love getting a piece of that action. When we offered the CFM56 to re-engine the air force's KC-135 tankers in the 1980s, they were happy to negotiate from the commercial market list price. They were less than happy when the bill for spare parts came, however. They thought they should be able to negotiate the same sort of parts prices they had on strictly military engines, in other words, minuscule or no profit on the sale of the engine and strictly limited profit on the sale of the parts. Just as we had with the issue of the refund on our calculation of inflation, we said that they could not have it both ways. (Again, a customer is a customer. The CFM56 core is based on the F101 engine. Wherever the replacement parts were the same, we agreed to give the air force the F101 price for the CFM56 parts.)

The program of re-engining the KC-135 has proven very beneficial to the U.S. Air Force as they now have 300-plus tankers with twice the productivity of the former tankers, with very little development cost and at a really good price for the engines. At the same time, GE and Snecma benefited from the base load it put into the CFM56 family.

The Great Engine War

In the late seventies, as the production of the J79 was slowing to a halt, it was apparent that, from the government sourcing point of view, GE was not a factor in the U.S. Air Force's fighter business, as all of their F-15s and F-16s had Pratt & Whitney engines. (We were on the navy's new lightweight fighter-attack aircraft, the F/A-18 Hornet, with our F404 engine, however.) Ed Woll, Morrie Zipkin, and Jim Krebs came up with the idea of putting the F404 and the F101 technology together. The F101-X, as it was initially called, was first funded by the U.S. Air Force as a demonstrator program, and we ran the first one in 1979. We eventually flew the engine on the F-14 in 1980 and on the F-16 in 1981. As a result, we had a tremendous amount of leverage in selling the engine program by our Washington office under the leadership of Harry LeVine and Cliff LaPlante. It also helped that the Government Armed Services Committee was getting more and more unhappy with Pratt & Whitney for not getting their F-100 fixed while the government seemed to be continually funding the program. It was not until we started to propose the F101DFE (Derivative Fighter Engine) that Pratt & Whitney started to react to the possibility that we might come up with an engine that could displace theirs. We had the Armed Services Committee visit our factory. We told them the advantages that the core would get from the F101, plus the experience we were getting from the commercial CFM56 development would make the F101DFE a really reliable engine. In addition, we promised that there would be no flight envelope restrictions on the planes with our new engine (a problem with the F100) as well as no reliability or durability problems. Our flight tests on both the F-16 and F-14 bore us out on these promises and gave us a great boost.

We also had some help from Pratt & Whitney who were more interested in stopping us from competing than fixing their own problems. As a result of the

Love and Hate—Working with the Military

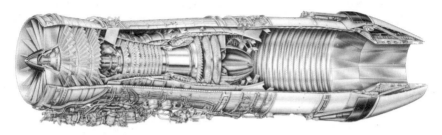

Cutaway drawing of the F110 engine.

pressure being put on them, they did recruit a few of our key engineers, Morrie Zipkin, Ben Koff, and Tom Hampton, who were very capable and knowledgeable. That really helped improve the F100, but by that time everyone in the government establishment was getting exasperated. A market comparison compiled by Krebs, Ward, and Harry LeVine really lighted a fire, and the U.S. Air Force started a competition that became known as the Great Engine War.

This specter of competition could have been just a government ploy to whip Pratt & Whitney into greater responsiveness, but people in the U.S. Air Force Procurement Office at Wright-Patterson in Dayton, Ohio, were also disenchanted with Pratt & Whitney. In February 1984, the government announced we had won a contract to power 120 F-16s compared to 40 for Pratt & Whitney. From then on the F110 (as the engine was now designated) won a major portion of the U.S. F-16 orders and was selected by many countries around the world. In addition, it was selected as the new replacement engine for the navy's F-14G Super Tomcat, which turned out to be a very successful program for both the navy and GE.

I was very proud of the military engine team under the leadership of Jim Krebs and George Ward, both of whom proved very inventive on the marketing side of the business over the long haul in getting this engine launched. They were continually visiting the Armed Services Committee, the U.S. Air Force, General Dynamics, and Grumman to tell them what our engine with its proven gas generator would do. They pushed and prodded, working all avenues, including the Israeli Air Force, so that the F110 was getting good marks all around. They never really knocked Pratt & Whitney—they did not have to. They just talked about what we were doing to deliver an improved product. Jim and George expended tremendous energy building confidence and developing strong relationships, and they contributed greatly to the total success we were experiencing as a business.

From 1984 to 1987, we ramped up to increase our total annual sales from just under $3 billion to more than $6 billion with big orders for the F110 plus orders for the F118 engine for the B-2 stealth bomber and the big success of the CFM56 and CF6 families. After we won some of these contracts, we had a very short schedule to get them into production. George Krall and his manufacturing team, plus all of the engineers associated with the buildup, really did a great job. This was doubly important with getting the F110 into smooth production as General Dynamics, who was building the F-16, already had a production line running with Pratt & Whitney's engines, and we could have caused serious second thoughts if we had missed a beat. This whole time frame was a great thrill for me. It showed just how well our team was working together with the various elements—the military, the airlines, and the airframers—all at the same time, and being successful in each area. It took a lot of work, but it sure was fun. Most importantly, the engines produced in that period for the military have proven very successful.

CHAPTER 6
Swords into Plowshares— the CFM56 Family

In 1969, just after we won the first contract with our CF6-50 on the Airbus A300, we began work on the concept for a new engine with around 22,000 lb of thrust. This would be a product needed to support an expanded demand for relatively small, single-aisle, twin-engined commercial aircraft that we saw coming. We had just won the B-1 bomber competition with our F101 engine, and the core of that engine looked like a good place to start, even though it had a relatively low pressure ratio for a high-bypass turbofan engine. We would have liked to have had a higher pressure ratio compressor to create more gas volume and velocity so that the low-pressure turbine could turn a big fan more easily. If we were starting from scratch, we would have probably designed it that way, but using the proven F101 core offered overwhelming advantages. For one thing, we could get by with a single-stage high-pressure turbine, reducing cost and complexity.

Snecma goes a'courting

The French government had been a major force behind the creation of Airbus as the European commercial airframe manufacturer. Now it wanted to exercise its support and influence to create a matching engine supplier. Government-owned engine maker Snecma had similar ambitions, and had begun conceptualizing their own "10-ton" engine, the M56. While Snecma's president, Rene Ravaud, dearly wanted a French entry in the commercial engine business, Snecma did not have market access that would help it sell a European engine in this thrust class, and neither, at the time, did it have the economic or technological strength to pull off such an ambitious task on its own. Snecma needed a partner with major commercial engine experience.

Ravaud and company went courting. They decided to hold a competition among GE, Pratt & Whitney, and Rolls-Royce for a likely groom to share the risk and initiate them as a player in the current world of commercial engines.

Rolls-Royce, the likely first choice as a European partner for a European engine, was still fighting its way out of receivership after the RB.211 debacle on the Lockheed L-1011, and so it was not much of a contender. In addition, Snecma had worked with Rolls-Royce on the engine for the supersonic Concorde, and the relationship had not been exactly cordial. There was not a great deal of hope that the non-existing friendship would turn to love.

Pratt & Whitney owned a 10% share of Snecma, and that gave them some influence with the French and the position of second most likely candidate. Even though Pratt & Whitney had proven difficult to deal with in the past, there were many at Snecma who were far more favorably disposed to working with them than with GE. Pratt & Whitney, however, was not particularly interested in going out of their way to make a new marriage. They already had an engine in the 20,000-lb thrust class, the JT8D, and while it was getting long in the tooth by now, the JT8D did have the pleasant circumstance of virtually monopolizing the market. Pratt & Whitney's follow-on spare parts business had made the JT8D a huge success and a proverbial cash cow—and they were not about to compete with themselves to upset that arrangement. They offered Snecma a joint-venture piece of a re-fanned JT8D in lieu of supporting the M56.

GE also had a history with Snecma. The relationship that developed with GE in putting the CF6-50 on the A300 had been both personally satisfying for Ravaud and financially rewarding for Snecma. As a result, GE had a well-deserved reputation for commitment and success, and we had the core engine technology that Snecma needed. Snecma had low-pressure system technology, government-supplied risk capital to develop it, and access to a Europe that had become territorially defensive about its markets. It sounded like it could be a marriage made in heaven.

GE proposed to make it a marriage of unprecedented equality. Unlike arrangements in which one company bought engine components from another at a specified price or those in which whatever profit was made would be split according to some formula, GE suggested a 50-50 revenue sharing deal. Each party was to design, develop, and manufacture engine components equivalent to 50% of the value of the engine. Half of the engines were to be assembled by Snecma and half by GE. Each party was to get half the price of every engine delivered, no matter who sold it or in what part of the world it was sold. Each party was also free to introduce cost cutting strategies into their own design and manufacturing processes to increase profits—one of GE's agendas—or use the business to grow a national technology base and support social labor programs—some of France's, and therefore Snecma's, agendas.

It was not that Snecma was unconcerned about making money or that GE was unconcerned about building technology or supporting social programs, but that the relative importance they placed on these issues, and many other issues, was different. By sharing revenue, and not profit, each party could introduce its own efficiencies and be true to its own mix of agendas. By sharing everything, instead of one party buying parts from another at a fixed price, not only would risk be shared equally, but so also would any success.

In the early 1970s, Ravaud pulled together a team to work with us to pre-negotiate an agreement that would establish the relative value of the various engine parts. Each company was to keep the actual cost of designing, developing and manufacturing the parts to itself. We reached a separate agreement on spare parts. Because GE was to produce the hot-gas-generating core engine, the part of the engine with the most frequent routine replacement of parts, we would logically be producing the lion's share of spares. Through a system of parts ratios, we created a way to offer Snecma a fair cut of the spare parts business.

The birth of the CFM56

The new engine was to be called the CFM56—CF from the GE tradition of using those letters to represent *commercial* turbo*fan* and M56 for Snecma's original designation for their engine. It would have a fan diameter of 62.5 in. and use the gas generator from the F101, which, because of its short compressor, made this a very compact engine. The new joint venture was to be called CFM International.

Because the gas generator was still highly classified, before we could get the program launched, we had to get approval from the U.S. government. It eventually took an agreement between President Nixon and President Pompidou of France before we could go ahead with the program. GE and Snecma had to agree that for three years, Snecma could not look into the design of the core (the gas generator) as it was still under USAF control. We also had to agree to a royalty payment to the government, because they had paid to develop the F101.

Our hope at GE was to use the experience of testing and operating the F101 on a production run of B-1 bombers to improve the durability and efficiency of the gas generator and make it more suitable for commercial service. Unfortunately, after Jimmy Carter came to the presidency, the B-1 bomber was canceled, and we had to add that unanticipated development cost to the ledger. Snecma was also finding the going to be more expensive than they had anticipated, but Rene Ravaud was always confident that we would sell the engine.

In search of a customer

We ran the engine on test in 1975, but we still did not have the glimmer of a customer in sight. The Airbus Jet 1 that was the anticipated target for Snecma's 10-ton engine was on permanent hold. The need for a fleet of smaller commercial aircraft that we had foreseen had not materialized yet in the eyes of the airlines. We were all dressed up with nowhere to go. We decided to go fishing.

The first expedition took us to T. Wilson of Boeing. Gerhard Neumann and Jack Parker had always been good friends with Wilson. Leveraging that relationship, they got Wilson to agree to fly four CFM56s on a 707 as a demonstrator. The combination was a really attractive looking airplane, but, although Boeing made a serious effort to sell the re-engined 707, their real focus was on getting the 767 launched, creating a new twin-engined airplane and making more derivatives of their bigger airplanes. The project went nowhere.

Next, we went to Jack McGowen, who had just formed CAMMACORP, which was to become a fine stable of engineers. McGowen, the former president of Douglas, had been there when the DC-8 was developed. The DC-8 was facing an FAA deadline for stage 3 noise regulation compliance. We worked with CAMMACORP to pull together a proposal to re-engine the stretched version of the DC-8 with CFM56s to meet the new requirements. This seemed to be our last chance to launch the engine. We felt we needed an order for 150 airplanes to make economic sense, but we would have lived with an order for 60 or 70 just to keep the program alive.

Pratt & Whitney was approaching the same market with an upgraded, but still relatively low bypass, JT8D. The CFM56 would offer more capability, be quieter, more fuel efficient, and more environmentally friendly; however, JT8s were everywhere. Pratt & Whitney's offering was established throughout the world with spare parts availability, and airlines would know what they were getting. Our engine was better, but it was new and untried, and we had to work to convince the potential customers—Delta, United, and the Flying Tiger Line—that we could deliver reliable engines on schedule under our CFMI production arrangements. McGowen worked out a great retrofit plan that would significantly improve these older airplanes beyond adding new engines, but we were not sure that that would be enough.

The day before the announcement was to be made, Fred MacFee, who headed the GE Aircraft Engine Group at the time, spoke ominously at a staff meeting, "If we don't get this program launched this time, it's canceled." Actually, neither GE nor Snecma wanted to be the one to cancel the program. Fallout from the French government over such a decision was expected to be severe. Plan B was really to put the program in cruise control, spending as little money as possible and waiting for another application to emerge. Luckily, our marketing team of Jack McGowen, Dick Smith, and Jean-Claude Malroux had done an outstanding job with these airlines, and the CFM56 was selected. From my perspective, while the relationship between Gerhard Neumann and Rene Ravaud was a very important prologue, this is the point where the real CFM56 story begins and the teamwork between GE and Snecma starts to pay off.

I, for one, felt a strong personal allegiance to the French team and their leaders, especially Ravaud. Unfortunately, Ravaud, who was a truly charismatic leader, was not to lead Snecma to the realization of the enormous sales he had predicted. Due to a conflict with the French government, he was forced to retire before the CFM56 could become the best selling engine in commercial aviation history, as it would do. He did succeed, however, in putting in place a showcase, modern engine assembly facility in Villaroche, France, capable of delivering the high production rate he had predicted. At this point in the story, though, while we could see possibilities, all we really had was this launching order to re-engine DC-8s, and so we put what marketing muscle we could muster toward selling the U.S. Air Force on using the CFM56 to re-engine their fleet of KC-135 tankers.

New engines for old—the KC-135 tanker

Originally, Boeing developed the 707 in the 1950s as a commercial airliner. Using the same general configuration but with a different fuselage diameter, Boeing sold it to the air force as the KC-135, a tanker used to refuel Strategic Air Command planes so that they could have extended range or stay on station longer. Many of these aircraft were in service at the time. Compared to the Pratt & Whitney JT3s that powered these planes, the CFM56s would mean a big increase in thrust. This would allow the tankers to keep pace with modern strategic aircraft. The CFM56 would also deliver a huge improvement in fuel economy, allowing a re-engined KC-135 either to fly twice as far on a mission or deliver twice as much fuel as before. In other words, by re-engining these airplanes, the air force would double their refueling capacity while allowing the airplanes being refueled to fly well above their stall speeds instead of just at the limit. Other valuable features of the new engines from a peacetime public relations standpoint were that they were quiet and they eliminated the trails of black smoke that came from the old engines on takeoff.

The fact that Wilson had worked with us to put four CFM56s on a 707 gave us a big leg up here. The people at Boeing had thought it was a questionable idea at the time, since the 707 airframe did not appear to have much residual life, but we eventually ended up re-engining more than 300 KC-135s.

The bid to re-engine the KC-135s put the Franco-American CFM56 in competition with Pratt & Whitney's all-American refurbished surplus JT3Ds, and, even though Pratt & Whitney helped fuel the official paranoia of depending on a foreign source for military hardware or sending U.S. tax money abroad, we won the competition. The engine we had designed for the air force's B-1 came back to benefit them in a way no one had ever expected.

Gambling on the Boeing 737

In the early 1980s, Ed Colodny, CEO of US Airways, outlined the requirements for a new medium-range 150-passenger airplane. We convinced Boeing that a new 737 powered by two CFM56s would be a good idea for US Airways and the marketplace. They agreed, but balked at changing the undercarriage height of the airplane to accommodate the engine's large diameter. We had already proposed a CFM56 with a smaller-diameter fan to Fokker for a new twin jet they were planning. Boeing said "give us that one." Ultimately, we changed the position of the gearbox from that of the original CFM56 and made the fan smaller so that the engine would fit—another new model.

There were many skeptics at GE. Creating a new engine version, even with a basic engine in hand, still costs considerable money. The earlier model of the 737, a small single-aisle airplane designed for short hops, had an inconsistent sales record and was a bit of a disappointment for Boeing. At GE, Harry Stonecipher and Ed Hood thought that, beyond being a gamble, putting an engine on the 737 was clearly a waste of money—but we went ahead anyway. The skepticism carried all the way to the top. In the first review I had with Jack Welch on this project, it was clear that our adventure on the 737 was considered quite a joke. No one thought we had much of a chance of convincing anyone to buy this allegedly old-fashioned airplane, new engines or not.

We hung in there, however, and our marketing people, along with Boeing, really pulled off a coup. US Airways played the lead role in launching the CFM56 on the 737, but the launching was also made with the help of Herb Kelleher, the dynamic boss of Southwest Airlines, who had a great relationship with Wilson. At that time, Southwest operated only inside Texas with a fleet of 23 aging 737-200s. Noise restrictions at their Love Field headquarters in Dallas were a major issue. Pratt & Whitney's refanned JT8D for the 737 did not resolve the noise problem, but the CFM56 did.

With this as a beginning and perhaps with more luck than anything we did, this airframe-engine combination proved to be phenomenally successful. The 737 went on to become enormously popular, with a number of new models of various sizes introduced, each powered exclusively by CFM56 engines. That first CFM56 for the 737 delivered about 20,000 lb of thrust. For subsequent 737 models, we developed a 24,000-lb version. The most recent version of the CFM56, for the latest 737, produces up to 28,000 lb.

In the end, GE, Snecma, and Boeing won big—and are still winning—with the 737. In retrospect, while I noted the skepticism of GE leadership at the beginning, it is perhaps more important to note not merely that we were right, but also that we were allowed the financial flexibility to proceed in spite of the skepticism and uncertainty. We had to prove our case, of course, but we were allowed to use our resources to modify and improve this engine for new applications well before the first model had been proven successful by turning anything like a profit. I should also add that, after their initial skepticism, Hood and Welch became great supporters of the program.

Airbus Industrie

At about the same time as the new 737-300 was launched, Airbus finally got serious about their twin-engined, single-aisle A320, the airplane concept that had convinced Snecma that they needed to design and build a new 24,000-lb thrust engine 10 years earlier. By now, however, the concept had grown a bit, and the original CFM56 that was powering stretched DC-8s and KC-135s would make the A320 a little underpowered. To be on the A320, Airbus wanted us to design a growth version of the original CFM56—another new engine.

From another quarter, Pratt & Whitney was beginning to see the handwriting on the wall. Their re-fanned JT8D did not even work as a stopgap against the CFM56, and, if they did not wish to concede the mid-thrust high-bypass segment of the engine market, they had better act fast. Proceeding as GE and Snecma had in forming CFMI, Pratt & Whitney and Rolls-Royce got together to form an international consortium to make a competing engine to challenge the CFM56. Joining these two major engine makers were MTU, Fiat, and Japanese Aero Engines with lesser shares in the new venture. They called their company International Aero Engines (IAE) and their engine entry for the A320 was to be the V2500.

We had hoped to convince Airbus to accept the CFM56 as is, but with the V2500 on the horizon, we had no choice but to bite the bullet, get out our checkbooks, and develop a third CFM56 model. The CFM56-5, as that new model was dubbed, was the launch engine for the A320 with sales to Air France and Lufthansa. This was a new market for Airbus, however; and they had a long struggle between winning the first orders and getting the airplane actually into production. After the enormously successful first test flight of the A320, the ever-prescient Rene Ravaud puffed his cigar and made a pronouncement. "We're going to produce 100 engines a month!" he declared. We all laughed, thinking he was making a joke of outrageous overstatement. He puffed silently for a few seconds, shook his head as the tittering stopped, then said quietly, "You young men, you don't understand. You don't understand." We came surprisingly close to that outrageous 100-a-month figure during many production years. Even in our wildest imaginings we never anticipated the success that actually occurred. Ravaud was certainly more accurate at prediction of sales than the rest of us were.

Squeezing out every drop

Airplane designers perennially seem to underestimate the power requirements for their designs. For one thing, they usually seem to be optimistic about the drag characteristics of their designs while they are still in the modeling stage. These are very complicated structures, of course; and, as good as computers are at prediction, the real proof is in the testing. The real airplane typically has more drag

Swords into Plowshares—the CFM56 Family

and is heavier than the designer's conception. Engine designers have learned to temper the airframers' optimism. If they say the airplane is going to weigh 800,000 lb, we say they will probably end up at least 10% more. From our standpoint, it is always good to have a lot of growth capacity up your sleeve as airplanes get bigger and heavier than the initial design.

By contrast with GE, the engineers at Snecma were always much more conservative than we were when it came to predicting how much performance we could get out of the CFM56. Neither GE nor Snecma thought that we could squeeze out more than 28,000 lb of thrust—that is until Airbus said they needed a powerplant for their four-engined long-range A340. Airbus was convinced that a gear-driven Super-Fan engine that was offered by IAE would give them what they wanted. We did not want to go anywhere near a gear-driven fan design because of the fact that gearboxes had always proven to be a weak link in any mechanical system. We were concerned, however, because Airbus was convinced that such a design would work. Eventually, Pratt & Whitney had second thoughts about the engine and retracted the IAE offer. This put Airbus in a bind. Lee Kapor and I went to see Jean Pierson, who was then the head of Airbus, and offered to build a conventional engine for him, and at the same time get our CF6 on the A330.

We began by proposing a 28,000-lb engine—which Airbus accepted—but it was obvious, even at the beginning that we would eventually have to come up with 31,500 lb as a minimum for the first airplane. We told our designers to design it that way, and they did. As soon as that first airplane flew, it was clear that 34,000 lb was a more accurate requirement, and that the airplane could make good use of 40,000 lb of thrust per engine, something that was beyond the CFM56 while still maintaining the fuel efficiency needed for a long-range airplane. The CFM56-5C was eventually launched on the Airbus A340 in 1987 at 34,000 lb of thrust. It remained the

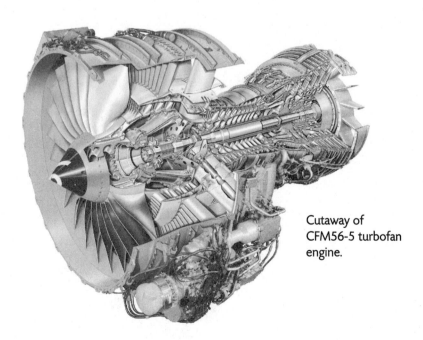

Cutaway of CFM56-5 turbofan engine.

exclusive powerplant on this very successful airplane, although a growth version of the A340 is now certified using a Rolls-Royce engine derivative.

Today, more than 14,000 CFM56 engines have been delivered, and there are 3000 more on firm order. Ravaud's imagination really paid off for both Snecma and GE.

Another important aspect of the program was the friendships we developed on both sides of the Atlantic. Many key contributors such as Pierre Alesi, Jean-Claude Malroux, Jean Sollier and Jean Bilien of Snecma, and Frank Homan, Mel Bobo, Frank Lenher, Ed Woll, Dick Smith, Bob Turnbull, Tom Brisken, and Mike Benzakein from GE have not only had outstanding working relationships but excellent personal ones as well. Ravaud did a great job in selecting his team, and that eventually led to other joint programs such as the UDF, CF6-80C, and GE90. After he resigned from Snecma, they had several changes in leaders, all of whom were government appointees, mostly from outside the industry. This meant that the GE team had to rebuild relationships with Snecma several times and, in some cases, help educate and persuade new leaders to support the CFM team agenda. Luckily, as I noted earlier, Ravaud had put in place some genuinely talented and dedicated people on the CFM team, making this effort considerably easier.

CHAPTER 7
Other Engines: M&I and UDF Non-Flying Gas Turbines

Not all of the products of GE Aircraft Engines were used to make things fly. In 1959, a version of the T58 helicopter engine was conceived to power hydrofoil boats and air-cushion vehicles and for generating emergency electrical power. It was dubbed the LM100 as it produced about 1000 hp for land and marine uses. This was closely followed by the LM1500, a 15,000-hp version of the J79 fighter engine, that was used to power navy gunboats and hydrofoils. (One of the first industrial applications was actually a ring of 10 J79s used to power a big turbine for Cincinnati Gas and Electric, a brainchild of Gene Steckley, one of the older guys. It is still used for auxiliary power in Cincinnati.) Converting these engines to stationary use did not take much modification from a mechanical standpoint, and so this was a fairly easy extra market opportunity. In the later half of the 1960s, the need for a more powerful marine powerplant just happened to fit with the development of the TF39 for the air force's C-5. The land and marine application of this engine was called the LM2500, with about 25,000 hp. The initial application was on the *Callaghan*, a navy cargo vessel that was already operating with two Pratt & Whitney FT4 engines. We proposed replacing one of those engines with the very first LM2500, and the navy accepted. After 2000 hours of operation in the salty air, the compressor and turbine blades were seriously corroded, but after developing special coating materials, everything worked fine. After 25,000 hours of successful service, we replaced the second FT4.

M&I Department

To a great extent, all of this was prelude, although the *Callaghan* experience gave us a lot of credibility with the navy. In 1968, GE decided to make a real business of turning aircraft gas turbine technology into something capable of being used in marine- and land-based environments, and created the Marine and Industrial Engine Department (M&I). Its first leader was Bob Miles. Bob was not an engineer, but he was Gerhard's confidant, human resources chief, and hatchet man. Gerhard gave him a pretty free rein, with the understanding that M&I would not divert any energy from the aircraft engines part of the business. This became the operating philosophy that functioned through my tenure as well.

The first major application possibility for the LM2500 was to be the navy's new 76,000-ton Spruance-class destroyers, with each ship powered by four engines. This was to be the biggest naval surface ship program since World War II and the largest single-ship propulsion contract in history. This was also going to be the base class for a number of follow-on applications such as hydrofoil missile boats, frigates, and the Aegis air defense cruisers, and so winning this competition was important. Sam Levine, the man I replaced when I went to Lynn, was put in charge of the LM2500 program. Our most likely competitor was General

GE's LM2500 engine.

Electric—the Steam Turbine Division. The corporate hierarchy had to make a decision which GE option to push. It all came down to which one would make the most money, and Sam, Gerhard, and Jack Parker made the case that the LM2500 could do that. There were some hard feelings between Power Systems and Aircraft Engines after we won this family feud. They lasted a long time. We went on to win the contract with Litton, the general contractor, and the navy. It amounted to 120 engines before we were through, the first mass production run for M&I. After establishing the engine with the U.S. Navy on the Spruance-class destroyers, we went on to power Perry-class frigates, Aegis guided missile cruisers, Kidd-class and Arleigh-Burke-class destroyers, and supply auxiliary ships. The LM2500 also powered ships of many other nations, including Australia, China, Denmark, Greece, Italy, India, Japan, Peru, Portugal, Saudi Arabia, South Korea, Spain, and Venezuela. It also was used for such things as providing power on off-shore oil rigs and pumping natural gas and oil through pipelines.

In the middle of the big production run, Sam Levine retired and Bud Bonner replaced him as head of M&I. Bud was quite a character. He had started in the navy as a seaman. After the navy, he opened a little marina in Florida somewhere, but began to be pressured by some very questionable customers with apparent underworld connections. He left Florida, got into the machining trade, and set up a small business. Eventually, he ended up as a machinist at GE. One day when Gerhard was walking the shop floor, he met Bud and was quite impressed. As a dirty-fingernails guy like himself, Gerhard took a shine to him. As a result, even though he had no college-level training, Bud moved through a series of ever higher-level jobs until he was promoted to be the M&I leader and ultimately became a vice president.

Bud built an outstanding reputation with the navy, and he really relished the strength of that relationship. He used to give talks at the Navy War College to admirals and other high-potential officers as if he were an ex-admiral himself instead of just a former swabbie. I think that gave him quite a thrill, as if he had pulled off one

Other Engines: M&I and UDF the Non-Flying Gas Turbines

of the world's greatest deceptions. Along that line, in the mid-1980s when he was getting close to retirement, Bud came to me and said, "Brian, my wife's not very well, so I think I'll retire early." A few years later, I learned that he was seriously ill. In retrospect, I think that it was his and not his wife's illness that prompted his early retirement, but he did not want anyone to know. I also did not find out until well after he retired that Bud's arms were covered with tattoos from his time in the navy. Even when we went golfing, he had always worn long-sleeved shirts.

Under Bud's leadership, we added the LM5000, with 50,000 hp, to our stable of engines. Using the gas generator from the CF6-50, the LM5000 provided electrical power generation. One interesting application was an LM5000 on a barge in Bangladesh. Cables came off the barge and provided power for the city. We sold several of these installations around the world. Bob Smuland followed Bud in the M&I leadership role. Work on the LM1600, based on the F404 engine, began under his guidance. Dennis Little succeeded Bob in 1989. Under Dennis's leadership, we added the LM6000, which used the CF6-80C2 gas generator. These people were really passionate. They absolutely loved the M&I business, and they did a good job at it. I did very little other than to approve their plans and ensure that they had a good support system and met their commitments. From my point of view as the chief, this was largely a business that ran itself. Although there were development costs and issues, they were generally those of ruggedizing and recalibrating to burn various types of fuel, at relatively low cost and low technological risk. In the field, there was not the same risk of product failure if the engine was sitting on the ground or was in a boat rather than seven miles up. We once had an oil-pipeline pumper that lost a turbine disk. That caused a bit of anxiety, but we turned it around very quickly and got it pumping again in short order. In general, these engines were incredibly reliable. One LM2500 ran for 50,000 hours—nearly six years—before needing an overhaul.

The ruffled feathers between Aircraft Engines and Power Systems were ultimately smoothed by splitting the M&I baby so that industrial applications for these engines fell under the aegis of Power Systems while Aircraft Engines retained the marine applications. This arrangement made a lot of sense, especially after Power Systems acquired Stewart and Stevenson, a significant customer of ours who packaged engines for industrial use and resold them. This was the ultimate way of convincing the power system guys that our gas generator technology could supplement their business rather than compete against it. Even though GE does not work on a commission basis for sale of these products, the Power Systems marketing people had always felt a lot better selling the products they were building in Schenectady, and sometimes, while we fussed with one another, Pratt & Whitney got the sale. The new arrangement, which occurred after my retirement and for which I take no credit, has proved to be quite successful.

Unducted Fan—the Ultra High Bypass Engine

In the late 1980s, the airline industry was faced with dramatically increased fuel prices. At the same time, we wanted to introduce some new technologies to improve our older engines. Harry Stonecipher and Bob Hawkins initiated a study of a very high bypass engine concept, which could reduce fuel consumption by at least 30% over existing engines.

The Unducted Fan engine being tested at Peebles, Ohio.

Art Adamson, who was always very creative, was put in charge of a small team to investigate this. After several studies it appeared we could get the desired reductions in fuel consumption by having a low tip-speed counter-rotating fan that would not only reduce fuel consumption but would minimize noise.

The question was this: how do we drive a large fan and at the same time keep engine weight low? There was consideration of a gear-driven fan, but remembering bad experiences with gear driven propellers, none of us really liked that solution. One of our senior engineers, K.O. Johnson, came up with the idea of a counter-rotating low-pressure turbine with the fan mounted on the outside of the engine. While this configuration was unconventional, it made a lot of sense from a weight and performance standpoint. It was also an opportunity to work on composite fan blades. We received a lot of encouragement from NASA, which also supplied some contract money. Using a F404 core, we built a demonstrator engine and ran it very successfully at the Peebles test facility. During an early test we lost a fan blade unexpectedly, which barely damaged the rest of the engine because of the low speeds. This gave us a lot of encouragement, and after fixing the blade, we ran an endurance test so we could flight test the engine. We flew on both the MD-80 and the B727, and the engine really looked promising.

As we were designing it, there were many skeptics, especially regarding noise, both interior and exterior. A group of Swedish scientists said it would really be noisy at altitude, but we proved them wrong. We also showed we could make an aircraft with these engines fly at the same speed as the current commercial engines. The interior noise was less than with Pratt & Whitney's JT8s. Of course a lot of people did not like the appearance of the engine with its big exterior fans on the airplane. We did start a new engine design with Snecma as our partner, but when fuel prices dropped we stopped the program to concentrate on already committed programs on CF6 and CFM56.

Art Adamson and his team received a well-deserved Collier Trophy for their efforts, and the technology of the composite fan blades was used in the GE90, so the program was not a mere design concept. In fact, I have no doubt that somewhere down the road we will take another look at this type of engine—both with and without shrouded fans.

CHAPTER 8
Leadership in a Large Technology Business

After the challenge of getting the commercial business launched and starting service in 1971 with the CF6-6 on the DC-10-10, the team of Hood and Rowe was dissolved as Ed Hood was promoted to group executive of the International Group. At the same time I was made a vice president and general manager of the commercial engine business. This was quite an honor for me—and the beginning of another great adventure. The week I was made vice president was quite an exciting one. I had just flown back from Europe to take a vacation on a chartered sailboat with my wife, Jill, our daughter, Linda, and our son, David. After being on the boat a couple days, I got a message from Gerhard to call his office. (There were no cell phones in those days.) We made an emergency landing somewhere in Maine. When I connected with Gerhard, he told me that one of our engines had lost its low-pressure turbine very close to the backyard of one of our best customers. Gerhard and I had to make an emergency visit to that customer in Los Angeles to convince him that we were on top of the problem and that it would never happen again. Feeling very responsible for the incident, I started the team at Evendale on an extensive program to find the cause and fix the problem pronto—which turned out to be tough to do. (We did it, though, with very little interruption in service. Our action ended up showing that we could fix problems fast.) In Los Angeles, the customer heard us out and seemed mollified. I returned to the boat, however, all my thoughts of a vice presidency had gone out the window.

A day after I returned to the boat, I received another urgent message to call Reg Jones, who was the chairman of GE at the time. I was in a bit of a panic as I placed the call, thinking that I was going to be fired. Instead, he told me that he was sure I would get the problem fixed and that I had just been elected a vice president. I was elated and rushed back to the boat to tell my wife, and we immediately set sail. About 2 hours out, my son, David, caught his finger in the door jam and crushed it so badly that we immediately came about and rushed him to shore for medical attention.

This left about two days of sailing before we had to get to Bangor, Maine and connect with the DC-10-10 that was just returning to the United States on the final legs of a world tour for a return flight to Cincinnati. Jack McGowan, the president of McDonnell Douglas, was on the flight as was the comedian Danny Kaye, who also was a pilot. All this concluded one very exciting week of vacation!

On my becoming a vice president, Harry Stonecipher was made the general manager of the CF6-6 program and John Pirtle the general manager of the CF6-50. Dick Smith became general manager of marketing. The certification of the CF6-6 took place in 1972 and the DC-10-30 went into service in 1973.

After a couple years, our customer base was growing fast, and the number of airplanes in the fleet was also growing—accompanied by the normal technical problems that demanded a lot of attention. It was decided that I should concen-

The father, son, and holy ghost—Fred MacFee, Gerhard Neumann, and me with Anna Chennault at a party for Gerhard.

trate on the marketing and Bob Goldsmith would head up the projects. While I did not like the idea at first, it certainly reduced my workload in Cincinnati. It also meant a lot more traveling, however, as we built up our customer base in the Far East. In 1975, when the airlines were going through one of their periodic crises, I traveled more than 100,000 miles and only sold one engine. This was very frustrating. All this traveling was unnerving my family and me, and in 1975, I suggested to Gerhard that, if Eddie Woll were going to retire, I would really like to run engineering for the whole business. This happened quicker than I expected, and Jim Worsham took over the commercial marketing job at the same time the airlines started buying again. We filled our order book, and I can only hope that all my traveling had some influence on the outcome.

My job in engineering was an exciting one. We were looking at a lot of new products at that time, and while it was not as time consuming as the jobs I had in helping create the re-entry into the commercial market, I was kept very busy working problems in both Lynn and Evendale with both the military and the commercial businesses.

In 1977, Gerhard Neumann had a heart attack and Free MacFee was put in charge of the group. Fred was stationed in Lynn and was also getting ready for retirement himself. He asked me if I would act without portfolio to coordinate the Evendale operation as well as do my engineering job. This was a great opportunity for me to understand the rest of the business better and work more closely with my peers—Ray Letts in manufacturing, Ralph Medros in quality, and, of course, the leaders of the commercial and military businesses.

Leadership in a Large Technology Business

In September 1979, I was called in to Fred MacFee's office and told he was going to retire along with Eddie Woll and Gerhard Neumann, who had been on semi-leave of absence. I was to get the top job in the Aircraft Engine Group. I was delighted. October 1979 was the official date for me to start the job. With the three top leaders retiring and Bob Goldsmith off to run a part of GE Power Systems, I had a bit task ahead to form a team. This gave me the opportunity to advance several people who had grown in the business with me. Harry Stonecipher was chosen to run commercial engines and Jim Krebs the military business. Frank Pickering assumed the engineering job. I was very lucky that Bob Desrochers stayed as CFO. This was a strong team that worked well together with the other incumbents. Shortly after I took over, Jim Worsham retired from GE and went to McDonnell Douglas and Ray Letts went to do a job in Fairfield. Ed Bavaria took Worsham's job in commercial marketing and George Krall took Letts' job in manufacturing. I had a great team. Another key move at that time was to ask Bob Turnbull to come from Lynn to head up advanced engineering. He later went on to do excellent work on the CFM56, especially in cost reduction on the gas generator that also directly benefited the F110.

I remained at the head of Aircraft Engines from 1979 to 1994. In 1994, I semi-retired and became chairman of Aircraft Engines so I could help Gene Murphy who had become president. This also enabled me to concentrate on helping the GE90 team get that engine through certification.

During my tenure as head of GE Aircraft Engines, the business grew from about $2 billion per year in sales to more than $8 billion. Although this happened on my watch, I did not make this happen alone. Neumann, MacFee, Hood, Parker, and Woll had built the strong team that I inherited in 1979. Others had the foresight to develop the GE1 building block engine in the late 1950s, creating the demonstrator engine that was to form the conceptual bedrock for every large engine we produced until the GE90. Still others had used their energy to create the alliance with Snecma that produced the CFM56 family.

While I had been a significant player on the leadership team since taking the job of general manager of the CF6 program in 1967—and a significant force behind the growth of the business—I had still been a player on someone else's team. My fellow players on that team were outstanding, as well. Surely, to paraphrase Isaac Newton, if I went farther in the days to come, it was because I stood on the shoulders of those giants who came before me and who worked beside me.

During my years at the helm, the business was financially and technically successful. To the extent that that reflects my leadership ability, I guess I was successful. I cannot say that I ever studied leadership, but I did watch others, and I did remember, and use, what I felt worked. Whatever intrinsic leadership talents I possessed that led to success in leading a multibillion dollar business, I am sure they are the same ones I used on the playing fields of my youth. As a result, any principles of leadership that I might divine now as my guides over the years will sound a bit concocted and simplistic in retrospect. In fact, the process of distilling any principles from my life is a bit foreign to me.

The actual process of managing was, for me, far more of an intuitive thing than a studied art. Engineering was always the foundation of my business life. On that foundation, I built a structure of genuinely liking people and liking to work with people. This led me to understand and appreciate strengths and weaknesses as well as needs and aversions. Always one to make a deal so that we could all play, I became very attuned to the marketing side of the business as well. I loved the engine business. I think customers sensed the enthusiasm that I had for the sport of making the world's best propulsion systems, and, unless the economic deck was unalterably stacked against us, they generally wanted to play the game with that enthusiasm on their side.

Picking people

The first principle of leading a winning team, of course, is picking the right people. To do that, you have to have some sense of who is available and who will likely be able to handle a new challenge well. Picking from inside the company is much easier than going outside for new talent. Inside GE, we have a good tracking system that identifies people with high potential, ensures that they are challenged by a variety of assignments under leaders with different styles, and predicts when they will be ready to grow into a new position. The GE system ensures that these high-potential people progress through a controlled program of training that takes them from their early days as a new hire to readiness for senior management. Through each phase, a leader looking to pick a new player for a position can ask people in the company, "How did that person do with this challenge?" or "How does he or she work in this kind of environment?" and know the meaning of the answers.

Picking people from the outside is much more difficult. Our human resources staff has developed very good interviewing techniques, of course, and anyone in a leadership position had to employ those techniques a number of times throughout his or her career. But no matter how good you are as an interviewer, you are never going to get the whole picture in an interview. Sometimes, you must depend on what people say about the person you are considering, and this becomes much more uncertain in this litigious environment in which an honest evaluation could result in a lawsuit for slander.

Prior success is one selection criterion, and, of course, you want to know if the prospective candidate works and plays well with others, but the real measures for a high-level player are skill and values. I always tried to pick people who were brighter than I was. This is not always easy for leaders, especially those who have even a dash of winning-by-intimidation in their character. It is critical on a high-performance team, however, for a leader to know his or her weaknesses, and to pick people for the team who make up for those weaknesses. For example, while I might have been a good marketer and a competent engineer, I was not particularly strong in financial management. Fortunately, my team included strong financial managers such as Bob Desrochers and Jim Park over the years. That is what it takes to excel as a business—a group of strong position players at the highest level who are also committed team players.

Unfortunately, egos sometimes get in the way, especially at the lofty levels where many of the players believe that fate has singled them out for greatness. Paradoxically, from what I have seen, the team members who spend more energy protecting their own positions than working more wholeheartedly for the team seem to hurt the progress of their GE careers.

To protect against that, I always tried to find someone who had advanced by giving his all to get previous jobs done. Typically, the kind of personal character that that takes is not going to change with a new title. I tried to avoid anyone who would look at a new job at a higher level as a promotion or reward instead of as a challenge. Rewards are what you get when you are finished; tougher challenges are what you get when you have proved yourself capable of completing difficult missions. I also found that people who moved relatively easily from one advancement to another seemed more inclined to coast rather than to pull hard. They often let you down when the going got tough. It was much better to have people whose mettle had been tempered by the struggle of hard work to overcome adversity.

I cannot underestimate the role of the person in charge of human resources (HR) in making the selection process both possible and painless for a leader. When I took over as head of Aircraft Engines, Don Lester was that man. Don was an outstanding evaluator of people. His groundwork in selecting only the very best people continued to bear fruit long after he had left that role. He helped give us some of the best people in the world to choose from. After Lester retired, I was assisted by several top notch HR leaders, Sam Dolfi, Chuck Chadwell, and in my last few years at the helm, Bill Conaty. Bill was excellent at his craft and went on to become Jack Welch's HR man.

Leading a team

In any event, no matter how you picked the players, you ended up with a group that was to be your team. As a team leader, I always found that it worked best to treat the team as a team rather than as subordinates to whom I doled out assignments and from whom I accepted reports. When we approached any problem, I tried to follow a few simple rules: find out what the team thinks instead of imposing on them what you think; listen fairly and openly to what they have to say; let the ideas percolate among the team until you, as the leader, can make a decision.

Because there are almost always a number of solutions to any problem, consensus in any dynamic field is usually not possible. If you are the leader, you have to decide. So, first, you must be open, honest, and genuine enough to inspire team members to create ideas that they are willing to defend passionately. Then, you must be perceptive enough to see through the passion to discover the idea that will work the best. Finally, you must be charismatic enough to get commitment from the whole team—defenders and detractors alike—to pursue the idea you think is best. This is not as easy as it all sounds, and many times it took a lot of restraint on my part not to bypass all of the democracy to get to a solution quicker and, in the process, to make decisions that were not democratic at all.

It takes quite an involved set of interpersonal communication skills to lead a team this way. Unfortunately, many people do not have those skills, but find themselves promoted into leadership positions anyway. Some of those leaders resort to fear as a motivator. Leaders who manage by fear do not have to develop any interpersonal skills whatsoever, but they do have to know more than anyone else about what is going on in every aspect of their operation. This is because people who are afraid of their boss either will not speak up when they see a problem or a potential solution or will tell him only what they know he wants to hear. The leader who leads by fear must know everybody's job as well as they do and keep accurate tabs on everyone through some intelligence network or inspection system. In short, leading by fear is a lot of work for the leader, it creates a lazy or an uncertain team, and it fails to produce a dependable flow of creative solutions—and usually ends up with a flawed operation.

In the past, you could see this leadership style in the bull-of-the-woods shop foreman who was promoted through the ranks. We would call people like them hardheaded when we were feeling kind.

In a way, Gerhard Neumann managed a bit like this. Although Gerhard had a great innate sense for leading and motivating people, he never built any close relationships with the members of his leadership team. Surprisingly, the people who performed best for him were usually the ones he seemed not to get along with well. I guess they were driven by pride to show him they were better than his opinion of them. His military background helped make him a militaristic leader, but he had an amazing personality that enticed people to work for him even if they did not appreciate his style.

Gerhard was a doer, a dirty-fingernails engineer with practical solutions. As a business leader, he hated to lose and was one of the most aggressive competitors I have ever met. Although he was not as analytically talented as a lot of the people around him, he was an outstanding field general. He led by example, and as a result, people would follow him as he charged the next hill even if they did not like him. Like IBM's watchword "Think," Gerhard's "Feel Insecure" sign seemed to epitomize his approach to leadership—never become complacent.

Applying Neumann's axioms

There were several traits that, looking back, I would classify as things-I-learned-from-Gerhard. The first of these is: don't delegate dirty work. By that I mean never ask someone else to do something you think is demeaning, embarrassing, or futile. Never tell someone to do something you would not personally be willing to do if you had the time, skill, and energy. Over the years, I worked for a few people who might ask you to try to sell something that everyone knew was wrong for the market or who would tell you to implement a policy that they had no intention of personally following or supporting. Gerhard was never that kind of leader.

Another thing that Gerhard did well was to stay close to the product, both as it was produced and as it was used. He was famous for walking through the production shops and the assembly lines and asking questions. He did something similar with customers, visiting them personally to see how our products were being used and what the problems were in maintaining them. When he went to Vietnam during the war, he talked with the servicemen maintaining our engines in the muck and dust of combat. When he returned, he led us to completely rewrite our technical manuals to reflect the reality of working on engines in field tents as opposed to in the modern maintenance facilities we envisioned when we initially wrote them.

Always an engineer at heart, Gerhard could not help himself from being intimately close to the product. Once when we were trying to win an order from United Airlines, he got up before their rather stiff executive team, spread out a bunch of drawings, and started explaining how everything would work. The United people were visibly taken aback to see a business leader acting that way, but I guess he inspired some confidence because we won that order.

On the downside, at home Gerhard frequently pitted the members of his staff one against the other. Not only was this distracting, but it tended to widen the apparent gap between engineering and manufacturing so that engineering seemed to throw theoretically buildable designs over the wall to manufacturing even though they could not be built with predictable results using the existing systems and equipment. To overcome the growing problem of interfunctional boundaries, Gerhard introduced the role of project manager throughout the business by 1966. The project manager worked directly to market a product line and had the referential power to coordinate the activities of engineering and manufacturing to deliver on promises to customers. This role went a long way toward solving the internal problems and had the added benefit of placing the emphasis of the business more directly on pleasing the customer. It was a good idea necessitated by what seemed to be a slip in leadership—and it took on a valuable life of its own.

In terms of relationships with the corporate headquarters, Gerhard acted at times as if Aircraft Engines was not part of the rest of GE. While he delivered the expected results, he seemed to treat GE's corporate headquarters as if it were a holding company and he ran an independent business that had nothing to do with other corporate ventures. He believed the GE hierarchy was too bureaucratic, but he also knew that he needed that bureaucracy to do some of the things he wanted to do. As long as the military was our primary customer, and the flow of funds into the GE coffer was positive and predictable, there was little interference. As soon as we started using large sums of money for commercial engine projects, the resentment over our somewhat arrogant isolation became apparent, and getting the funds required a harder fight than it really should have.

Gerhard especially resisted sending any of our senior managers to GE's training center in Crotonville. Perhaps this was because he did not want anyone in the rest of the company to see how good our people were for fear they would recruit them into other GE businesses. In any event, he only sent the people he wanted to get rid of, which sometimes worked. This standoffish spirit of *not invented here*, whether initiated by Gerhard or not, seemed to be part of his legacy. I think the remnants of that attitude often hurt us.

In any event, when I took over his former job as head of Aircraft Engines, Gerhard gave me a lot of invaluable advice about working with commercial and military customers, and the intricacies of dealing with the corporate headquarters in Fairfield. He tempered these things with three pieces of memorable advice. The first thing he said was, "Brian, it's very easy to neglect your family. Don't do it." I think that was an aspect of his life he felt he had missed by paying too much attention to the game of business. Second, he said, "Remember that the customer is the boss." I always tried to live by that advice, and I think it made our business better even when I found myself defending a customer's position against increasing our income. Finally, Gerhard advised, "Never let them talk you out of the company plane." In his last few years, he did not have full use of a plane and often found himself on commercial flights, sitting in airports and waiting for luggage. He had to retire early for health reasons, and there is no question in my mind that the wear and tear of commercial travel shortened his years at the helm.

Compared to Gerhard, I think I had better vision on where to take the commercial engine business than he did. Gerhard was used to getting paid up front by the military and making fixes with their money, and the risks of the commercial market seemed to frighten him. I saw that, if we were to remain a contender in the engine business, we had to have a strong commercial presence. There was really no choice.

Gerhard and I also had some essential style differences. He hated confrontation. On the other hand, I liked people to argue with me—and argued with Gerhard all of the time, much to many people's dismay. Gerhard hardly ever fired anyone. If it had to be done, he would get someone else to do it. While the worst part of life in any business is to call someone in and say, "You're not cutting the mustard, and you've got to move on," I felt, if that was my decision, I should break the news.

Gerhard was a hard guy to get close to, while I had social relationships with everyone on my team. We simply had different personalities. His was large and towered over the business. One of the hardest parts of my taking over, even though it was a year after his retirement, was getting people to believe that Gerhard was not still there any longer and that I was running the show.

Learning from Jack Welch

Jack Welch would become another of my great leadership influences. I never really knew him until he became the chairman. Jack grew up in GE's plastics business, which was new-venture technology at the time. As a hot shot in the plastics business, Jack was a clear contender to replace Reginald Jones when it was time for Jones to step down as GE's chairman. Unfortunately, there were other crown princes who were also in contention: Ed Hood, Tom Vanderslice, John Burlingame, and Stan Gault. For more than a year, the organization had to put up with these people, who were all good, each acting as if he could walk on water and trying to show he was better at it than the others. Vanderslice and Gault were particularly competitive, and people down the line in the organization had to watch everything they did or said because they did not know which of the princes would become the next king.

At Aircraft Engines, we were all pulling for Ed Hood—not that we had a vote. Ed had been one of us, and we wanted to have a friend making decisions about the future of our business. When I met Ed in 1967, I knew him as an outstanding thermodynamicist, but over the years in GE's International and Power Systems businesses, he had become a really impressive business leader as well. After Jack was selected, Ed went on to be vice chairman and Jack's technical advisor, but he never seemed to be quite at home at the headquarters in Fairfield. I would speak to him nearly every day in that role. He acted as my sounding board and, I think, saved my bacon a number of times by keeping my concerns to himself rather than passing them on to Jack.

In the process of committing to new leadership, I am not sure how Reg Jones decided on Jack or why it took so long, but I think it is clear he made the right choice. Jack had been a hard charger at Plastics, and in a world where things were

Leadership in a Large Technology Business

With Jack Welch, center, left, and GE Vice Chairman Paulo Fresco at the 1989 Paris Air Show.

moving fast, he was especially aware of the bureaucracy that was bogging GE down. I believe Reg saw the need to tighten the organization, and Jack was probably the one person among the contenders who would move the mountains necessary to get that done.

Moving those mountains earned him the behind-his-back title of Neutron Jack—the man who left buildings standing but, like a neutron bomb, eliminated the people. It is true that at the beginning of his tenure many bureaucrats bit the dust, and if operations did not perform, Jack had no mercy with them. The result, however, was a huge business with vast resources operating with the speed and creative energy of a bunch of small, entrepreneurial businesses. Not only did Jack do great things for GE, but his example also led the whole of industry to do great things.

He was certainly right in the observation that bureaucracy is the killer of initiative. When I took over the Aircraft Engine Group in 1979, just before Jack became CEO, we were a $1.8 billion business, and we had 550 executive-level people. We cut the number by 20% to 440. As a result, we got rid of a lot of meetings but did not hurt the business at all. In fact, by the time the business grew to $8 billion on my watch, we had hardly risen from that 440 number. Getting more with less was definitely a Jack Welch trait.

Another thing I admired about him was his fantastic memory. As the leader of General Electric, he had to deal with 13 highly diverse businesses, with 13 very aggressive, dynamic leaders in each. Many commitments of resources and promises of results were made by word of mouth only. Jack showed an amazing ability to keep all of this straight. He was not someone to whom you said something and hoped later, when things were not going as well as you had planned, that he had forgotten.

I always found it very easy to speak my mind with Jack, and although a number of people would get annoyed with me when I took him on now and then, I truly believe that is what he wanted. I operated as if I were the expert in my business, and while he might be able to bring insight to what I was up to, it did not serve GE for him to get deeply involved in the day-to-day operations of my business.

There is no question that, as he became more successful, Jack took on the qualities of a legend, and people became more intimidated by him. In many ways, I think that adversely affected the entrepreneurial spirit that was the hallmark of the early Welch years. On the other hand, Jack always felt I was more interested in making good engines than in making money—and I guess he was right in that. In any event, I must have made enough good decisions and Aircraft Engines must have made enough money during my years at the helm for him to have put up with me in spite of some of the disputes we had, or I am sure he would have replaced me with someone else.

In general, Jack and his corporate management team listened to the entreaties of the business leaders as they put in their bids for a share of the resources at GE's disposal, and depending on the passion of those entreaties, he usually acquiesced even if he was not fully convinced. As I mentioned earlier, in 1981 I pled my case for launching the CFM56 on the Boeing 737. It would mean decreasing the size of the fan so that the engine could fit under the 737's wings, and neither Jack nor Ed Hood, who managed the technology sector that the Aircraft Engine Group was under at the time, agreed. They accepted our proposal, albeit with great reservations, and when initial orders were slow in coming, I received constant harassment from both of them for a few months. It turned out, of course, that the CFM56 on the 737 became a better engine program than any of us could have imagined. Jack could have stopped it, but he trusted my conviction—and gave the project his full support—even though he was not personally convinced. He held me accountable, of course, but he gave Aircraft Engines a chance to prove what we believed. That approach was characteristic of his leadership style and something I tried to emulate.

I think I can credit Jack with bringing Aircraft Engines into the real world of making money by leading us to always do better. When his tenure began, we thought our lives were about making jet engines and making them safe. As long as we made more profit than Pratt & Whitney, which was relatively easy since Pratt & Whitney was a notoriously bad performer, we were happy. Jack set the bar much higher for us, and he taught us to love it. I remember the thrill the whole executive team got the first time we delivered more than a billion dollars in operating margin. Jack showed us that you could have as much fun making money as you can have making engines.

Inspired by Rene Ravaud

Another leader who was a great inspiration for me was Rene Ravaud. Rene was the head of the French engine maker, Snecma, when we began the CFM56 program with them. The French are known for their stubborn nationalistic pride, and Snecma was a nationalized industry on top of that. While it seemed very un-French of him, Rene saw GE Aircraft Engines as having developed a knowledge base that Snecma was lacking and could put to good use. In a brilliant move to maximize the value of our partnership, he sent some of Snecma's best people who were particularly fluent in English to be posted in the United States as part of the new organization we had formed to develop, build, and sell the CFM56. Those people cemented what was to become a wonderful relationship, and they

Rene Ravaud, Sir Frank Whittle, Brian Rowe, and Jack Parker at the GE Propulsion Hall of Fame.

became personal conduits of invaluable technical understanding between GE and Snecma.

No complex relationship is ever totally sweet, however. Over the years, Rene and I had many disagreements, especially about cost, price, and marketing strategy. Many a meeting ended with him chomping fiercely on his cigar and me red in the face, but in the end, he worked harder than anyone to protect the team.

In times when our leaders in Fairfield had doubts that the CFM56 program would ever sell more than a couple hundred engines over its lifetime, Rene was inspiring the team with, "We'll build 100 engines a month." And eventually we succeeded. His belief in what could be attained by our partnership and his knack for picking just the right people created an enormous fuel of inspiration that helped pull us through many crises. There have been many programs of international cooperation over the years, such as for the Concorde and the Tornado. Each of them seemed fraught with suspicion and backbiting. CFM International was different, and Rene Ravaud made it so.

Unfortunately, leadership of a French nationalized company is a political job. Rene fell out of favor with a new administration and was replaced. Jacques Benichou, his successor, was an excellent manager. He knew the industry very well, and he worked hard to keep the CFMI team successful. He soon retired, however, and was replaced by a series of people from government, who understood neither the aircraft engine business nor how a profit-making industry worked. We soon understood how important leadership was in making international cooperative ventures work.

Direction, consistency, respect, and communication

As I said before, my leadership style is not something truly studied. I have been blessed by having many good people to watch—people like Gerhard, Jack, and Rene. What developed as my style was what I picked up from watching people like them and using what I thought would work for me. As a result of thinking about it now, some clear lessons come to mind. I think that the most important lesson I learned was that, as a leader, I had to provide a sense of direction for everyone on the team. Perhaps the best example of this occurred when we eventually became number one in new engine sales. I was concerned we would get complacent, and I started the Continuous Improvement Program to drive the team to focus on doing even better. (I have more to say about this program in Chapter 10.)

A second lesson was consistency. A leader needs to be a Rock of Gibraltar, not only visible, but also solid and unwavering. Nothing is worse than having a leader who changes his or her mind about the team's direction so frequently that everyone holds their fire to avoid either wasting effort or making what the leader will consider mistakes. When I stood up for my belief that we had to change the diameter of the fan of the CFM56-5 if we wanted to stay a contender on the 737, or when I vigorously supported the CF34 for the regional jets, the Aircraft Engines team knew I would not wither under pressure from the corporate headquarters. That made us strong in our resolve.

Third, while it is important to be able to joke among members of the team, it is critical that the joking never demeans anyone or undercuts someone's position with sarcasm. Sometimes the leader has to act like a referee to keep everyone playing fair. In my experience, the potential for this kind of mean-spiritedness occurred frequently between Manufacturing and Engineering. We often had to walk a fine line between the need to ship the product on time and the need to put in a fix before shipping it. These situations were ripe for critical exchanges in which each side would characterize the other as inane. As leader, I had to teach the team—by example mostly—to honor the disagreement without disparaging their teammates.

Fourth, a good leader must be a good communicator. I have known a number of people who were so convinced that knowledge is power, that they were reluctant to share it. In my experience, however, I have seen that when leaders share all of the knowledge available to them, the team takes on a sense of power and an enthusiasm that energizes both the team and the leader to be and do more. It also opens the floodgate to ideas from below. Trusting relationships and open communication are the first things that suffer when leadership begins to fail. Wherever people are reluctant to speak up to defend what they care about, it is a good sign that the leader has become a bureaucrat, and the team is stagnant. Finally, a successful leader must have a passion for the product, the company, and the team—and not be afraid to show it.

While I wish I had done all of these things perfectly all of the time, of course I did not. In fact, sometimes I was too stubborn in getting my point of view across. I also could be rightly criticized for being too soft on people who had served us well in the past but who were not up to par in their new assignments. I guess that part of being human is to fall short now and then, even while knowing what to do.

The GE management system

A leader is someone who sets his own goals to grow the business using the available resources better than his boss or the directors think he can. As such, leadership is one aspect of running a business successfully. Another is management.

The manager takes those goals and applies the formal system by which the business makes decisions and takes action to successfully reach them. Any head of a business is both a leader and a manager. While leadership may be a unique, personal thing, management is something that must operate within a shared system that everyone understands. I am quite lucky to have worked within GE's.

To convey the goals of leadership at GE, each year began with a general managers' meeting. This meeting brought together about 500 executives from GE's 13 businesses in Boca Raton, Florida, to listen to significant advances in individual businesses that could apply to all of the businesses. These leaders also met each other on an informal basis and got a direct look at the chairman's vision and goals. This session was the jolt that jump-started each new year. Each business came home with videotaped excerpts of the chairman's message so that we could preach the true gospel with conviction and get everyone marching toward the same objectives. Those objectives always included meeting our financial goals for the current year and revisiting our long-term plans so that we could continue growing the business.

I liked to have the Aircraft Engines leadership team meet in sessions we called Time to Think, or TTT. In these sessions, we would set aside our daily routines and focus exclusively on developing a flexible plan to meet the goals of the organization for growth and sustainability, to meet the goals of the customers for safe, useful, and cost-effective products, and, most importantly, to meet the financial goals of the company. (In general, we tended to shoot for 17% or more operating margin and 10% or more return on sales to meet our financial commitments as one of GE's mature core businesses.) In this process, we had to analyze the cost, timing, and projected payoff of each engine project we wanted to push. Obviously, there would never be enough resources to do everything we might want to do or to compete on every possible aircraft application. There was often as much crystal ball gazing as there was science in some of these projections, but we would make them with conviction and then do our damnedest to fulfill our own prophesies.

In March, before GE issued its first quarterly report, each business leader and his finance manager went to Crotonville to swap ideas and to report on our individual progress toward meeting the year's goals. Jack would ask incisive questions and, more importantly, remember the answers—and the commitments.

Finance managers at each business reported up the financial chain of command, and they were directly rated on how well they delivered the numbers. Of course, so was the leader of each business. Working together as a team was very important. In truth, I always did well working in these dotted-line reporting situations because of my reliance on the team to help make rational decisions rather than only counting on my intuition to chart the course and charisma to get people to follow trustingly.

Preparation for these quarterly Corporate Executive Council sessions could get increasingly intense, especially as we approached the fourth quarter and the time to make any course corrections was slipping away. The last thing we wanted to do was miss on delivering on our numbers, and so at the end of each quarter, and especially at the end of the year, there was a mad scramble to ship engines so that we could book them as sold and reduce our inventory figures.

By the time we got to the end of the year, we had sometimes turned off our suppliers for a month or more to meet our inventory numbers and were working an enormous amount of overtime to meet our projected delivery schedule. Usually, we had squeezed the supply and production channels so hard by the end of the year that we were incapable of assembling another complete engine until mid-January, and we laid off the same assemblers we had been working overtime until 31 December. This was an area of frustration that seems to be prevalent in the whole industry, and we still need to improve, even though the introduction of Continuous Improvement and Six Sigma methodologies did take out much of the variation and flatten our delivery schedules.

On the surface it may seem that yearly financial goals that can cause such quarter-end and year-end distortion in the system are disruptive—and they are—but the situation is really more like a runner sprinting for the finish line or a professional American football team performing with renewed intensity in the last two minutes of the game. The truth is, we loved the game of making our commitments, and the role of leadership was to make everyone involved enjoy the game and not feel like galley slaves with the drummer continuously increasing the beat.

In fact, we did care about people, but we cared most about people who were gamers, whether that was as specialists or as leaders. In April or May, Jack and his corporate human resources people would visit us to review the individual performance of about 300 of our top managers and hear our plans for their further development. This is something we called Session C. GE's strength is built on having the very best people, and Session C was the culmination of our program for personnel growth. Every person of high potential was systematically reviewed in terms of strengths and weaknesses, career challenges, and learning experiences. Training and education at Crotonville as well as specific engineering, financial management, and marketing development programs were available for people who demonstrated ability, energy, and leadership, and Session C kept us on our toes making sure we were developing the people to whom the current leadership could pass the torch. I have subsequently found that a strong bench saves in recruitment costs as well as helps a business continue without missing a beat if one of the top people leaves.

In June or July, we would go to Fairfield for Session I, a full-day presentation of our latest three-year strategic plan. This is the plan we had nurtured in our TTT session. It took into account our view of the limitations on business resources—in other words, what projects might deliver the highest return in the shortest time with the least risk. Strategic planning is an almost impossible job for companies that make aircraft engines and the aircraft they power because of the long development cycles for those products and the relatively high dollar-volume of products that have to be sold before those development costs can be recovered.

Typically, the cost to design and develop an entirely new commercial engine is about $1.5 billion over four or five years, not a commitment a business would enter into casually. The risk is so high, that unless a major customer agrees to purchase that engine for a number of airplanes when it is available, we would start dragging our feet on the development or stop entirely to avoid hemorrhaging money. Assuming we had a major launch customer and, with luck, several smaller ones, we would then complete the job of designing, building, testing and certifying the engine. Then we could begin production. Of course, since some of the material needed to build jet engines must be ordered a year or longer in advance, we needed to begin procuring production materials and hardware long before the development engines had been certified. In other words, we had to bet that everything would work at least well enough so that we would not have purchased material that we could not use. Getting the timing right is excruciatingly difficult.

The launch customer knows that there is at least some risk involved in trusting us to meet our commitments. To accept that risk, the launch customer negotiates a lower price, one that may cover the cost of building and supporting that customer's engines but that does little or nothing to offset development costs much less deliver a profit to GE for its huge investment.

In fact, a given engine line may not begin to break even for 10 years after the first production engine is delivered. Some take even longer. While we project the needs of the aerospace industry well into the future—so many aircraft of such and such a size over the next 20 years with so and so percentage of the market going to GE—no one is bound by our predictions. The truth is that beyond three years, the predictions are pretty much guesses, sort of like the weatherman's prediction for the weather a week from now. (Planning does become a bit easier the more engines you get into the field. Because most airplanes stay in service 25–30 years, you can do fairly realistic planning for your spare parts and service load. This has become much more complicated in recent years, however, due to unforeseen events such as the 9/11 tragedy.)

In the midst of this situation, we needed to deliver on our financial commitments to the corporation every year. To do that, we had to be the industry's lowest-cost producer so that we could win sales with competitive pricing while still delivering significantly better returns than our competition.

Flexibility, the drive to win, and the love of the game still are the strongest characteristics we look for as we develop new leaders. And the game is making money. A Boeing executive once said of us, "Pratt & Whitney loves to make engines; GE loves to make money." In a way, that could be seen as a comment negating our appreciation for the wonders of technology in overcoming the limitations of gravity. I look at it in a different way. To make money in this business, we must not only please customers with timely delivery of the world's most effective and most dependable jet engines—at a life-cycle cost that is second to none in the industry—we must also do it with an internal efficiency that allows us to make significantly more income from each sales dollar than our competition. While that is not exactly the same as blazing new trails of technological possibilities, it still needs an exciting and demanding game plan. That is the plan we take to corporate headquarters in Session I.

Throughout October and November, the chairman and his team spend a day with the leadership from each of the businesses to follow up on the personnel issues that were raised in Session C, to get a report on how things are going with this year's budget, and to review how the long-term goals from Session I will be accomplished. This meeting is called Session II.

Also in October, about 140 of GE's top executives meet at Crotonville to help develop the strategy for the following year. This meeting lays the groundwork for the meeting at Boca Raton in January that establishes those new goals. And so, the yearly cycle repeats.

Taken together, the GE management system works constantly to renew the company, to raise the performance bar higher and set new business goals, and to develop and evaluate leaders as they strive to meet those goals as inventively as possible. Working as a leader in that environment is a daily challenge. There is no room to coast after a success, hardly time to take a breath, but success in that pressure-cooker can be invigorating. You either love it or you hate it. If you do not love it and try to run with it anyway, the machine will grind you into the dust the first time you stumble. But if you love the personal challenge of putting yourself on the line every day to reach the next business goal, this is the most invigorating team sport that was ever invented.

The only way a Jack Welch or anyone else could manage a large company such as GE is by having truly dedicated leadership in each of the businesses—leaders who not only have good leadership skills themselves but also have strong people around them who understand the products and will stand up for them. It is very important that the leader of the company has a powerful and consistent memory backed by a management process that is built on consistent financial and human resource measurement systems. Mistakes will always be made, and it is key that a leader learns from them but does not dwell on them. Unlike batters in baseball, in business, one's success ratio can never be in the 30% range. It must be well above 80% if the leader is to maintain credibility with both the people in the business and the stock market.

CHAPTER 9
Manufacturing—The Hidden Heroes

Most people associate me with the design engineering part of the business; however, I have always had a deep connection with and affection for manufacturing. Without the manufacturing people we could not deliver anything. Not only could we not make the production engines, but neither could we make the initial development engines. The contribution of manufacturing to the development of the jet engine was much more than simply doing what the designers described. Materials engineers had to develop materials that were tougher and stronger than anything anyone had imagined. Then manufacturing engineers had to find ways to cut, drill, and machine these most extreme materials to make the parts the design engineers conceived. This not only took incredible focus and dedication, but it also involved considerable financial risk. To make some design concepts work efficiently and cost effectively over the long haul, we knew we would need novel materials and processes, but usually no one could predict with any certainty whether an engine in development would see extensive production. Spending money and energy to develop materials, manufacturing processes, and tooling that would only pay off after a long production run was a risk that had to be weighed carefully. Luckily for us, working together, the design engineers, the materials labs, and the manufacturing people came up with the right answers more often than not.

Early lessons

My appreciation for manufacturing formed when I was an apprentice. I learned the fundamentals of manufacturing—filing, drilling, milling, sawing—and spent many hours operating machine tools that I learned to respect. I remember grinding some flanges on a big grinding machine. Inadvertently, I forgot to switch on the magnetic chuck, and as this big grinding wheel passed over the parts, these flanges let go and started whistling around the room. Luckily none of them were damaged, and nobody got hurt, but that taught me a lesson—to respect the methodical setup and operations procedures for these machines. I also came to the realization that it takes a lot of know-how to define the planning for these complicated machines and, in some sense, a lot of courage to work with these powerful tools.

One of the most dislikable jobs I ever had as an apprentice was rolling threads on the studs that held the engines together. This was certainly a task that called for full automation, but in those days we did not have the computer support we have today. As a result, this job was done by hand-feeding blanks into a thread rolling machine and then gauging the results coming out. This really intensified my desire to improve my education so that I could be a design engineer rather than spending my life as a machine operator. As another result, I have a great sympathy for machine operators. Some of these jobs are really boring, and it is no wonder they try to find ways to get around the tedium by doing things that really test the patience of their managers.

When I was about 18 or 19, I was asked to do a layout of a production facility for making the main drive shafts that connected the turbine and the compressor of the Ghost engine, which was in production at that time. The manager for production was a very nice, highly educated, sophisticated man who was absolutely useless in getting the employees to do any work. The smartest thing he did was to employ a very tough red-headed Irishman who used every word available in the English language to motivate his people. There was no such thing as Continuous Improvement in those days, and a tongue lashing was the order of the day for any failure. Again, this inspired a sympathy in me that carried forward to people in manufacturing I interacted with later in my career. When I was finally able to do something about it, I tried to get more input from them as to what would work and how we could improve our production processes.

The production line at deHavilland, while not big, did produce some pretty good engines. Interestingly, all of the people who ran the shop—the foreman and the superintendent of the shop—except for the really upper management, were all shop-trained people. When they had to handle some labor dispute, the situation was often a bit of a conflict for them because many of them were more sympathetic with the people they were controlling than with management. DeHavilland was not a very big place, and we used to have continual visits from upper management walking the shop and talking with the people. This conveyed a sense that the leaders understood and appreciated what the workers were trying to achieve. When I got to GE, I discovered a situation that was quite different. Most of the people running the shops had never operated the machines and had never experienced the tedium of some of these operations. As a result, GE had a different sort of relationship between management and the labor.

Materials, processes, and people

My first opportunity to work with manufacturing at GE came after I had been there about a year and a half. I was assigned to produce some designs for the lift fan. Assigned to me were a value engineer, a quality engineer, a materials engineer, and a purchasing engineer. Manufacturing's objective was to make parts as simply as possible using standard procedures. Quality's objective was to make them inspectable. (They would prefer to inspect a cube rather than a complicated shape.) The materials engineer preferred you used something off-the-shelf, and the purchasing agent hoped it would be cheap. Everybody had his agenda and his own function to optimize. I was charged with coordinating all of this. (As we matured, manufacturing became more and more responsible for quality. As a result, we now get a much better integrated design at higher quality levels.)

As part of this project, I took a trip to visit a supplier in Dallas, Texas, who was making one of the parts for the lift fan. It was really the first business trip I took for GE. I went to the Cincinnati airport, got on a DC-6, and flew to Dallas, where I was met by the Ling-Temco people. They took me to see the part, and it looked terrific. It was made out of material 0.010-in. thick, and so it was almost like paper.

As a designer, when you draw these things, you get one feeling, but when you see them being made, you get a completely different impression. The part was made of 17-7 PH material that was able to be worked pretty easily while it was being made. Then it had to be heat treated to give it strength. So here was this really beautiful part, about 60 in. in diameter, very thin, and they put it in the heat-treat oven. It came out with very little distortion, which pleased us, and all was going along very well until a worker accidentally knocked the part on the floor—and it broke into a million pieces. Obviously it had been heat treated too much. You can imagine the devastating look in our eyes when we saw this part, with all of the hours that had been spent to produce it, laying there like pieces of broken glass. This taught us an expensive lesson that this material was not suitable for use until we could come up with a different heat-treat regime, one that gave it a combination of toughness and flexibility as well as being strong and hard.

The people who contributed greatly toward creating innovative materials and the processes for machining them came from the Thompson Lab in Lynn and the materials and manufacturing labs in Cincinnati. They had to develop materials that would withstand very high temperatures at high stress levels while being crack tolerant. Their materials had to pass not only the test of strength but also one of machinability. They helped us produce precision cast turbine buckets, and they came up with electrostream drilling, laser drilling, and many other very inventive ideas for cutting and shaping the strong, tough materials needed to operate at extreme temperatures. These materials were considerably stronger and tougher than anything we used to machine in my days as an apprentice.

Laser drilling of a turbine airfoil.

Another interesting innovation developed by GE was the method for precision rolling small compressor airfoil shapes. We did this in our Rutland, Vermont, factory, but the whole process was developed at the Thompson Lab in Lynn under the guidance of many great materials engineers. These very small compressor blades are the size of razor blades. To do them by any other process was extremely expensive, and to get the quality we required was very difficult. GE's foresight in creating processes like pinch rolling helped us reduce cost in our engines before our competitors could.

When I was made manager of the engine design for the CF700, I was very lucky to have Phil Whitmore as my manufacturing manager. He was responsible for coordinating the requirements for the engine with manufacturing. Of course the CF700 was quite a small program within Lynn at that time because the J85, T58, and T64 were in full production, but we had a very good development manufacturing shop under the leadership of Paul Brigham. Our engineers spent a lot of time with his people, not only making sure they interpreted the drawings correctly but also seeing that we were being charged properly. This might seem strange because these were development parts, but even in the development phase we needed to watch the budget carefully. I had an interesting financial and program control manager working for me named Ernie Haas. He used to keep meticulous records of how much we were charged every month. We used to question the engineers as to why so much was being spent on particular parts because it frequently looked like it should have been a lot less. I used to walk the shop every day and I could never see the number of people working on our parts that we were being charged for. After several heated discussions we got this all sorted out, but it took constant vigilance to ensure our parts were being charged properly. Frankly, we did not worry too much about other people's parts being overcharged; all we wanted was to be sure we were not being charged for work done on their parts. As a result of all this, I developed an unfortunate reputation for being hard on manufacturing. That was never really my attitude because I always knew perfectly well that without their cooperation we could never do the job. It was true, however, that the newest program in the shop usually was charged the most unless someone was really, really watchful.

The manufacturing process is not simple. A jet engine has about 30,000 parts, and all of these parts have to come together in final assembly at about the same time. This timing issue is particularly pressing for development engines. Most of these 30,000 parts are made by outside suppliers. In fact, about 60% of a GE engine is procured from outside the company. Our engineers had to become familiar with not only our manufacturing techniques but also those of our suppliers. The quality standards of our suppliers were critical, and they took a lot of monitoring. Sometimes that monitoring was not all that it needed to be. In one instance, I was having a part made by a supplier in Portland, Maine. A quality engineer was assigned to go to Maine week after week to monitor these parts. Frankly, I couldn't understand the necessity because the part was nowhere near completion. Out of curiosity one day I went to Portland to see this part being made, and I asked the person running the shop, "Where is our quality engineer?" He replied, "Well, he comes in on Monday, looks at the part and the progress, comes back on Thursday or Friday, looks at it again and goes home Friday afternoon." I asked, "Well, what does he do the rest of the week?" His "I don't know" was apparently an attempt to cover for the chap. It turned out he used to go fishing in the lakes of Maine while he was supposed to be working. Either he was not as enthusiastic about this program as the rest of us or no one had told him what to do. I learned from this that a critical aspect of managing is ensuring that people know exactly what they have to do and when they have to do it. Of course, the manufacturing organization was really unhappy when I came back and told them this. They said, here is Brian Rowe trying to show us up again and telling us how to do our job. That was not my intent. I just wanted to make sure we weren't being charged for something we should not have been charged for.

As a result of these sorts of experiences, in my projects we worked very hard at protecting ourselves from being overcharged. When I got to Evendale and was running the CF6 program, which was much bigger, this took a lot more vigilance.

The process of going back and forth between the design engineers and the manufacturing engineers really helped GE Aircraft Engines get its costs down. It was always our objective to beat our cost targets, particularly on the military programs. The government accounting processes put some very tough targets on us every year, and our objective was to beat them so that we could make a little more money. They knew more about our cost system than we did, and so it was a constant struggle to come up with ideas to improve our processes to reduce the cost without putting in a lot of risk.

Even though we were pretty sophisticated in the design of this equipment, we did have many many *costly reductions,* as we called them. These are instances in which people have modified a part slightly for improved productivity only to find that when it gets into service that small modification did not work quite as well as the original. A classic example of this was a J85 compressor blade that was being made by an English company. We felt we could produce it less expensively, and we pulled it back into our factory. We made it exactly to the drawing, but our parts never worked as well as the ones made by the English company. To this day, we still do not understand. There must have been some slight difference in the dimensions, and the interaction between the aerodynamics and the aeromechanics design of the part came into play. This conundrum notwithstanding, in general we did an outstanding job with airfoils. All airfoils have to have precisely rounded leading edges to assure there are no stress concentrations. Many of the excellent manufacturing techniques that we developed helped us do that efficiently, and we eventually produced blades at about one-third the cost of our competitors.

We were also blessed with many great manufacturing leaders for whom I had much respect. I think the most impressive manufacturing feat I witnessed in my career was George Krall and his team, including Chuck Chadwell and Dick Burk, gearing up from a relatively low production rate of $2.3 billion of sales in 1983 to $8 billion in sales in three years. While they never did get to the point where everything was delivered in accordance with a uniform, flat-loaded schedule, they always seemed to make their production commitments by the last day of the month or the last week of the quarter. This always caused a great deal of anxiety—not only for me but also for the people in Fairfield who were counting on us producing the projected number of engines—but I can say that George and his team always came through. In spite of the anxiety, I never really doubted that he would do it.

The whole idea of assembly in an aircraft engine factory is much different than in an automobile factory because of the number of units produced. Generally, we were lucky to produce 300 engines of a given design in a year. Of some engines, like the gas generator for the CFM56, we produced nearly 1000 a year, and so we were really able to get volume-related cost reductions there. Most cost reductions, however, came from improved techniques. I would like to suggest that the path to those cost reductions was always clear, but that was not always the case. In one instance, the value engineers were considering a cost saving on certain brackets on an engine. They came in one day and said, "Hey, Brian, we can get this part for 10% less if we make it a fabricated part rather than a forged part." We redesigned the part as a fabricated part, and just as we were ready to go into production, the forging supplier discovered—under the incentive of losing the business—that he could deliver the part as a forging for less than the cost of a fabrication. We discovered that if we jumped back and forth trying to get the lowest cost, it really forced our suppliers to be creative in getting their costs down.

Special processes

With some of the very high strength—and hard to machine—materials that we used, we were always looking for ways of forging the parts closer to the final dimensions we wanted in the part while being sure we did not get any cracking or impurities in the material. The inspection process on the turbine wheels and compressor disks that were made of titanium or very high strength alloys really cost a lot of money. The forging for the turbine wheel of the GE90 cost on the order of $15,000–$20,000, and so it was imperative that a forging that expensive not become scrap because of a mistake in the manufacturing process. We had to be very careful. Some of the material we were working with was barely cuttable with conventional tools, and we developed ceramic tools for cutting these very high strength turbine alloys efficiently.

Another area manufacturing pursued with the materials lab was making parts out of powdered metal. This process was originally conceived in the 1960s by the Materials Lab under Lou Jahnke. Rene' 95, a proprietary alloy, was the chosen metal. The theory was to make an extremely fine powder, contain it in a mold that was near the final shape of the part, and put the container under high temperature and explosive pressure. A near-net-shape casting would result, saving a considerable amount of money. This is true in theory, but that first part of the process became the toughest part—that of getting an extremely fine, pure powder with the alloy combined properly. While we got this process down to a fine art, it took us 5 or 10 years longer than we ever dreamed it would. I remember we used to bedazzle the airline guys with the process when we demonstrated how we used it on small parts. It really was difficult to get good turbine discs and shafts when we began, but now this is a well-used process that produces excellent parts.

Another process that proved very effective for us was something initially developed at Caterpillar. This was the process of inertia welding. This process joins two pieces of what would become a rotating assembly in the engine. One part is fixtured securely; the other part is attached to a massive flywheel and spun. When the two parts are brought together, the inertia of the flywheel is converted

Inertia welding.

into the heat of friction, and the parts are precisely welded. I think GE was the first to use this process in aircraft engines. It allowed us to join parts that, because of their shapes, could not be joined in any other way and, because of their size and contour, could not be easily made as one piece. Inertia welding enabled us to get very good and strong welds. Without all of these techniques, the cost of the jet engine, while very high today, would be prohibitive.

It took dedication from people in manufacturing to come up with these ideas, and a lot of courage on the part of designers to use these processes in the design of their parts. I remember that our competition used to say about our film-cooled turbine blades, "These blades are going to fail. They'll never be any good. They won't last." Of course they lasted longer because they had this nice film of cool air surrounding the turbine bucket while it was operating in a nearly 3000°F gas. Our manufacturing people had to learn not only how to drill the holes but also how to get rounding, or chamfers, on the edges to reduce stress concentrations. The design engineers had to learn to design the leading edges and where to put all of these holes so that the blade could operate without excessive internal stress. Although our competitors told everybody that these blades were useless, after we ran our first engine with these parts, it was interesting to see how quickly they introduced the use of this concept.

While outside suppliers contributed significantly to the manufacturing progress we made, I was frequently upset by the fact that they always seemed to be relearning lessons on our time or money. This was particularly true of the ball bearing vendors and the people who produced raw materials. I expected that, if they had fixed a problem they were having with a bearing or a material they were making for one of our competitors, they would come to us and tell us—and be prepared to apply that fix to our parts or materials. Instead, when we found a problem in the engine program and went back to the supplier, they would say, "Oh, yeah. We had that problem before with a Pratt & Whitney part or Rolls-Royce part, and this is how we fixed it." It seemed to me they were double dipping all of us. (I may not be fair on this, but this is how I felt.) Another issue I had with suppliers was that, if there was a failure of a supplied part because of a quality issue and the failure caused the whole engine to fail, costing us a lot of money, they would never really contribute to the solving of the problem. I realize many of them did not have the resources of GE, but it was upsetting because they just stood and watched. We set up many vendors—not only in electrostream drilling but also in many other areas of manufacturing—and they took our technology, produced parts for other companies, and made a profit from it. We were hog-tied because there were fewer and fewer vendors who wanted to participate in this very technical and expensive business, and it is getting worse as we go down the road.

Staying number one

In the 1990s, just after we reached the status of being the number one engine producer, I was worried about how we would keep the enthusiasm and energy going so that we could be number one forever. Being a sports enthusiast, I noted that teams that become champions usually sink very rapidly to being number two or three—and even to the bottom of the league sometimes. I knew that one way to keep the energy was to rearrange people; however, a change of attitude would be much more constructive. Our Continuous Improvement Program helped do just that. Our quality manager, Jim Nelson, showed me a video of W. Edwards Deming talking to people at Ford. I thought if Ford can do it, we can do it. We brought in some Deming-inspired consultants, and they really helped build teamwork, even though sometimes they seemed more socialistic than I thought they should have been. It is true that we needed to do a better job of working with our employees—we needed to give them freedom of thought—but we could never guarantee them lifetime employment. The market controlled how many people we could employ without committing economic suicide, and it was necessary to be very careful about setting reasonable expectations. Deming's position was that, as quality improved and fewer people were necessary to inspect for and fix mistakes, those people could be set to work on new ventures. While this philosophy may have some credence in an expanding economy, it does not work in down cycles. (I say much more about Continuous Improvement in the next chapter.)

We were lucky to have been able to begin our Continuous Improvement program during an economic upswing. We needed to free as many people as possible from unproductive work just to meet our commitments to our customers, and Continuous Improvement became a rallying point that kept us focused on staying number one in the industry. When the eventual downturns came, the disciplines of Continuous Improvement and Six Sigma, which by then had become a way of life, allowed us to hold our own with fewer resources than our competition.

Manufacturing—The Hidden Heroes

Manufacturing really led this effort as it spread throughout the business, and there are many positive things that manufacturing taught the rest of us about Continuous Improvement and Total Quality. Manufacturing people are really the hidden heroes of this industry. Design engineers get most of the recognition for creativity, and the finance people get the kudos for the monetary results, but if the people in manufacturing had not produced the parts on schedule to exceedingly high quality standards, none of us would be working. In this business, as in any business, it is the team that makes things happen. Yes, there must be good leadership, and there might even be stars, but the willingness of the team to work closely together to achieve the mission is what really counts. Manufacturing is the bedrock of that team. I am proud of what they did for me.

CHAPTER 10
Lighting the Torch of Quality

Quality has always been essential to our product of engines for airplanes, and that quality has always been rigidly controlled. In fact, our product made us passionate about quality. We knew we could not let 400 people get on an airplane and hope we had done the right thing. The weight of responsibility for lives was just too heavy to be uncertain. So it was not as if, sometime in the early 1980s, we suddenly discovered that having quality products would be a good idea. We already knew that, and we were already delivering the best quality that money could buy. In fact, that was the problem. Our approach to quality was to throw money at it. We would make some parts, and our inspectors would inspect them. Good parts would go in one pile, and failing parts would either be reworked, scrapped, or inspected again by the customer and a team of engineering experts to see if they could be accepted as-is.

The good parts would be combined into subassemblies, which would be inspected and tested. Again, the good ones would go on; the failing subassemblies would be taken apart and fixed. Good subassemblies were joined together to become engine components that were inspected and tested. This process went on until we had a complete engine that had passed its own inspection and testing—and was ready to be shipped to the customer. The process worked to deliver a quality product, but it was expensive, uncertain, and slow.

This kind of quality system developed over time, and while it was far from perfect, it was the best thing we had. The machines we had available to make the parts did not deliver the accuracy or the repeatability to produce consistent results. Machine operators would fiddle with their machines to keep them on track, which resulted in either making things worse or hiding the scope of our control problems. On top of this, the military, who were the initial customers for our products—and were still about half our business in the 1980s—loved inspections. Our agreements with them specified not only how their product would be made and how it would perform, but also how often and at what point in each process it would be inspected.

Within the ranks of hourly employees, an inspector's position was relatively lofty. It usually involved no hard physical work. You simply measured something or verified something that some other skilled worker had done, and you certified that it had been done properly. It was a little like being the umpire at a ball game. The players did the work, and you got to say if everything was acceptable while they waited for your decision. Inspectors liked their jobs, and being made an inspector was usually a promotion.

A culture committed to inspection

All considered, there was a strong cultural bias toward doing an ever better job of inspecting quality into our products. The new idea that was to be the seed crystal of the quality paradigm shift was that we should be able to control our processes so well that we consistently produced products that were so good that they never needed inspection. Clearly, selling that idea would take a massive culture change, both internally and externally.

The whole concept of inspection is based on separation—that someone outside of the person who made something must verify it. As a result, we had a separate Quality department that was like a shadow to our Production Division. In fact, we would have to overcome functional specialization throughout our organization if we wanted to dramatically improve our approach to quality.

Compartmentalization

In a way, we were a lot closer to a total-quality environment in the early days of the aerospace business. Then, cross-functional teamwork was a natural part of getting things done. Engineers followed the parts they had designed as they went through development and into production, and we were in touch with the customer throughout the whole process.

As the job got bigger and more complex, we moved from integration of the various functions—engineering, material sourcing, production, sales, finance, customer service—into separation and differentiation. We did this because it was the best way we could think of to keep people motivated and to stay competitive. As a result, however, we had become a group of specialized experts, and in many ways, each expert's horizon had shortened to the arm's reach of his or her specialty.

Along with this myopia came bureaucracy. Departments and divisions erected fences to keep myopic specialists from bumping into things in their environments, and these fences, which may have helped them, became road hazards to the rest of the organization. Each function defended "their way" as the only way they could work, and soon we found ourselves translating to each other rather than speaking the same language. The system itself generated differing goals. As a result, there were misunderstandings and conflicts. Things slowed down. When we forced them to speed up, every aspect of the business got more expensive.

On the other hand, while all of this was bothersome, the system seemed to be working. We were close to an $8 billion business with about a 60% share of the large commercial engine business. We were the number one business in our market. We were on top of the world, but the world was burning—and we had seen what it had done to the once dominant American steel and automobile industries. Even though our business was not officially broken, a few of us were convinced we needed to fix it. Passing that conviction to the people who would bear the day-to-day work and upheaval of change would take some doing, however.

Benchmarking

First, we looked at some of those really efficient companies everyone was talking about to see how they operated. After some consideration, we decided to follow the path that had been philosophically blazed by W. Edwards Deming, the man credited as the quality-savior of Japan who advocated the incremental evolu-

Lighting the Torch of Quality

tion of quality through the understanding and improvement of processes—rather than by beating up people who made defects. We hired a guru and a team of Deming-driven consultants and set to work turning our world upside down. We called what we were doing Continuous Improvement.

While other companies typically had taken a slow approach to learning total quality and evolving, we felt that, building on their experience—and being GE—we could do better. Our culture was based on know-it-all, do-it-all managers, and so naturally we thought we could change faster than anyone had before. We started with a vision, of course: build the best product today and tomorrow; be the share leader in the marketplace; make our customers ecstatic with our products and services; deliver lots of money toward corporate income; have the happiest, most energetic employees in the world; etc.—in other words, the same vision we always had but which had never been on a plaque on the wall before.

Top down

To reach toward that vision, we developed a two-pronged enveloping strategy—top-down and bottom-up—designed to meet with a bang at our middle managers, the people who were already doing their utmost to keep our inspection-based system from going totally out of control. The top-down aspect began with the senior staff. We agreed to commit two full days each month as a group to our own training and education on total quality issues and techniques. One of the most important things that we learned in this process was how to listen. Some of our leaders had a tendency to tell people how to solve problems rather than to listen to what they had to say. Even when a democratic boss speaks first, ideas stop flowing. Our teachers and consultants trained us and gave us procedures and admonished us when we slipped to ensure that listening became a part of what was to many of us a foreign lifestyle. I must admit this was hard for all of us, particularly me.

Out of this grew the Nine Initiatives, a collection of loosely defined but far-reaching missions. The Nine Initiatives were apportioned one to each staff member. That staff member was to apply the total quality techniques we were learning and, by involving his staff, to pass on the techniques and values of Continuous Improvement.

The quests that the staff had were missions such as speed up the engine development cycle, create common goals between design and manufacturing, reduce inventory and increase the inventory turn rates, reduce dimensional variation in manufactured parts, and improve the quality of our delivered products.

Because we were sending our senior staff out like bishops to convert the multitudes, we created a clergy to accompany them and do the work of ministering and proselytizing. We trained this cadre in every aspect of quality improvement and dubbed them Total Quality Advisors.

Bottom up

Our Continuous Improvement initiatives were running in parallel with an overall GE effort called Workout. Workout was developed at the GE corporate headquarters and embraced by Jack Welch as a way to attack inefficient processes in every part of the business—the factory, the office, customer sites—and to identify unnecessary work that could be eliminated. In practice, a Workout session usually began with something like a town meeting in which everyone involved in a process, from the department manager to every individual worker, met in a room. As in a town meeting, each voice was to be heard as equal to any other voice. The Workout session began with people expressing opinions about what could be done better. Groups formed to examine ideas that showed promise. By the end of the Workout, normally a two- to three-day process, the groups came back to the town meeting with recommendations. The manager was required to approve each recommendation on the spot, reject it with justification, or, if further research was necessary, to consider it at a follow-up meeting.

For many employees, this was their first chance to have a voice in structuring their own work life, and the energy of that empowerment helped fuel the bottom-up aspect of our intended quality revolution. It also reenergized many old labor-management conflicts. Some unions decided their members would not participate in Continuous Improvement projects because they were not being paid to improve the company as part of their agreement. Others in the organization wanted a stipulation that no one would ever lose a job as a result of the efficiencies we hoped to derive from Total Quality. Of course, we could not make that promise.

Consultants

When we began our journey into Continuous Improvement, we employed the help of consultants in the Deming mold. Deming's philosophy was that the efficiencies generated through improved quality should produce increased opportunities for workers. Japan, Inc., was still a proponent of employment for life at that time, and our consultants believed that if Deming could do that for Japan, they could do that for America. Unfortunately, we hit a dramatic decline in demand for engines in the middle of this process, and leaders must lead. Over our consultant's objections, we let people go. In fact, since Continuous Improvement had made us more efficient, we were able to maintain a scaled-down operation with fewer people than we otherwise would have needed. This allowed us to deliver positive operating income when the rest of the aerospace industry was in deep financial distress. The layoffs did end much of the euphoria associated with the quality movement at Aircraft Engines, and we and our consultants soon parted company.

Continuous Improvement vs Workout

Both Workout and Continuous Improvement had the same general objectives: to use teams throughout the business to simplify work and make it reliably repeatable. Many perceived a battle of orthodoxy, however, between "Jack Welch's plan" and "the Aircraft Engines' agenda" for quality improvement. Some half-expected a corporate inquisition to descend on us at any moment, forcing the vice presidents to recant their connections to Continuous Improvement and embrace Workout while burning the Total Quality Advisors at the stake for their arrogance.

Lighting the Torch of Quality

I often told Jack that eventually he would have to embrace Continuous Improvement, but he just replied that I did not understand his plan and implied that we were elitists. A cartoon that we used at the time was a parody of the "Less filling!" "Tastes great!" Miller Lite beer ads. In the cartoon, the two sides were shouting "Continuous Improvement!" and "Workout!" at each other as if there was substantial difference between the two. Jack and I did have our differences about how to approach quality, but we each had the same objective in mind. I think that Jack was trying to get the whole GE company started on a road of cultural change that would lead to the places many of us at Aircraft Engines had already passed through. Like the adventurous children on a hike, we were unwilling to wait for the leader and went exploring on our own, coming back with reports of what awaited ahead.

By the way, GE's current business-wide Six Sigma methodology for quality looks much like our statistically based Continuous Improvement did, and today's Black Belts and Master Black Belts bear some resemblance to our Total Quality Advisors. Six Sigma is a more inclusive, better integrated set of tools and processes, but I like to think that we at Aircraft Engines started the quality ball rolling for GE, and that the leader had to keep up with us adventurous kids.

Everybody's someone's supplier

Our first task was to make quality personal. Serving your customer is a lot like loving your neighbor. Just as you might be prompted to ask, "But who is my neighbor?" we wanted people to be asking, "Who is my customer?" Answers such as "my boss" or "the end user" indicated an impersonal connection to quality. We wanted people to see the next person who got their work as their customer—and that their job was to serve that customer as well as they could. While this sounds self-evident and logical, it was not always how we were doing business.

For example, traditionally the Engineering Division designed engine parts to work in theory. (In truth, they also focused on producibility, but operability was the first consideration.) After they proved their concepts with a few prototypes, they sent the designs to Manufacturing. When Manufacturing got them they might say, "We can't make the things you've designed efficiently. In fact, some features we can't make at all." Engineering might then feel justified in replying, "Proving the design is our job; making the production model is your job." It would often take some design changes and qualifying new manufacturing processes before the perfect design became an engine able to be manufactured perfectly and at a cost we could afford. After some time working with Continuous Improvement and Six Sigma tools, parts and whole engines are designed today so that they can be manufactured and maintained with existing equipment and skills to deliver virtually perfect products from the very first day of production.

As another example, let's say you supply A, B, and C. The next operation after yours puts together parts A, B, and C to make component D. Once you get rolling, it's more cost-effective for you to do long production runs. In the past, at the end of week one, you might ship 100 of part A. Week two it's 100 of part B. Week three, it's 100 of part C. Meanwhile, your customer, the next operation, has to own and store 100 As and 100 Bs before he can get to work. At the end of week three, he puts his whole crew on overtime to meet schedule. Your costs go down while his skyrocket. Now, we would employ synchronous-flow production, making parts in the smallest lot sizes possible so that every step of the operation can be sensitive to demand at the next level of production—all the way to the final customer. None of this could have happened unless people saw that the next operation was their real customer.

Modules and commodities

As we focused on the needs of the customer rather than the marginal efficiency of our facilities, we began to see new ways to pull work together into modules. In the past, if the operational steps to make a part required milling, drilling, turning and grinding, we would first send all of the pieces to the department with all of the milling machines. Then they would be packaged in some fashion and sent to the drilling department, and so on. If a lot of people in the drilling department were on vacation, our parts might sit there for a while. Sooner or later, the parts would be finished and get back to us. Today, we would put a milling machine, a drill, a lathe, and a grinder in one location. We would teach all of the workers to operate all of the machines. We would develop a continuous flow so that every so often one part would be completed and go off to our customer—located right next to our manufacturing cell.

Just as we modularized on a small scale, we took the same idea to a larger scale by organizing around engine commodities. Jet engines are made of components that can be lumped into families that have similar design, manufacturing, and support requirements. For example, turbine airfoils are the bucket-shaped blades that catch the hot gas flow coming out of the combustor and, by spinning the high- and low-pressure turbines, turn the compressor and the fan. As I described earlier, these blades operate in an environment that is hotter than their melting point. They are cooled by casting them with hollow passages inside, drilling hundreds of tiny holes into those passages with lasers and blowing air from the compressor through the blade and over its surface. Designing and making turbine airfoils that will work efficiently and last for a long time is at the cutting edge of our technology.

We decided to take commodities such as turbine airfoils and combine all aspects of their design, manufacture, and support under one organization we called a Center of Excellence (COE). The Turbine Airfoils Center of Excellence, for example, would have designers, sourcing and purchasing people, manufacturing resources, finance people and customer support people—everything necessary to support that commodity from cradle to grave—reporting directly to the manager of that center. While this had the drawback of creating a new kind of specialization, it had a big offsetting advantage of ensuring that the heritage of specialized technology was being maintained. In addition, as designers walked the shop and visited customers, they got first-hand feedback from both the manufacturing arena and the field. The result was cutting-edge technology and cost-effective delivery.

Today, there are Centers of Excellence for fan and compressor airfoils, rotating parts, structures, combustion, turbine airfoils, configurations, controls, and

Lighting the Torch of Quality

product testing. The COE system has many advantages from the standpoint of cost, quality, and technology development, but it does isolate these commodities in many ways. As a result, it is imperative to have good systems designers and managers to pull it all together. This need demanded a personnel development program that lets people bridge the disciplines of these centers.

Six Sigma Quality

As Continuous Improvement got rolling, our next step from the basic tools of quality improvement was a concentrated effort essentially to eliminate all defects in everything we do from design through manufacturing to administration. The program was called Six Sigma Quality. It had been developed and used with great success by Motorola.

The logic behind Six Sigma Quality is that, no matter how well you control them, all processes vary. When the variation is too great, the result is a defect. For example, if you were typing and hit one key slightly off-center or with a slight variation in pressure, you would probably still type the word correctly. With a little more variation, you would hit a wrong key or not depress the right key far enough. The result would be a misspelling—a defect. Six Sigma methodology would design the keyboard-operator interface so that the result would be fewer than 3.4 defects per million keystrokes. In other words, you would type about 500 double-spaced pages—a good-sized book—before making one mistake. That is Six Sigma Quality—virtual perfection. Six Sigma Quality puts inspectors out of business. Human inspectors simply would not be able to find defects from a Six Sigma process.

We started employing and adapting Motorola's Six Sigma techniques at Aircraft Engines just before I retired. I am pleased that Jack Welch, with the help of Gary Reiner, his Corporate Quality guru, changed his initial opinion of Six Sigma and spread it throughout all of GE.

Chapter 11
Overcoming a Failure of Integrity

The F-16 with our F110 engines won orders in many countries around the world. One that gave us great satisfaction—because it was where our engines would be put to the toughest test—was Israel. This order, however, also gave me and the rest of us at GE one of our greatest setbacks from a moral point of view. The person in charge of procurement of engines for Israel was Brigadier General Rami Dotan, who was considered a great patriot by the U.S. Air Force and the Israeli government. Unfortunately, he was dishonest, and his dealings hurt many fine people both in the Israeli Air Force and at GE.

The Dotan affair

Dotan was a very forceful character. After one lunch I had with him at the request of our sales representative covering Israel, Herb Steindler, who turned out to be his co-conspirator, my wife said that Dotan was one of the scariest men she had ever met. How she came to this conclusion I am not sure, but her feelings were certainly a valid premonition. Dotan and Steindler, together with several other Israelis based in the United States and Switzerland, came up with a scheme whereby they were misdirecting U.S. military funding into their personal bank accounts. The money was being released through GE as the contract administrator to the Israelis for flight testing and for the building of some engine test cells and other tooling.

In May of 1992, one of the people in our Military Engine Operations received a call from General Dotan threatening to cancel Israel's contract with GE for F110 engines because a GE employee was trying to gain unauthorized access to a classified airbase in Israel. The GE man was just trying to confirm that the last of five test cells was complete before payment was authorized. To placate Dotan, we asked our man to stop trying to physically inspect the test cell and to use other means of documentation to verify its completion. It turned out that this was a mistake. There was no test cell being built, and this inspection could have brought this scheme to a dead halt a few months earlier than actually occurred.

At about the same time, Dotan was conspiring to eliminate one of his Israeli accomplices. The person he asked to do the deed informed the Israeli authorities, and that blew open the whole scheme. Our people, we found out in the investigation that followed, had been unwitting accomplices by accepting false affidavits claiming that work was performed on these test cells so that the Israeli Air Force (Dotan) could be paid. (Dotan was also double billing both GE and the Israeli Government for the flight-testing program. This was a considerably larger fraud in terms of the dollars involved than the test cell scam.)

All of this was expedited by great pressure from Steindler—in the name of the customer. Our people were naïve enough to believe the work was being done, and clearly we did not have good checks and balances in place. In addition, our main technical representative in Israel, rather than letting us know what was going on, had been collecting evidence on his own behalf so that he could file a government-backed qui tam (whistle-blower) suit against GE. This could make a very exciting novel or a movie, but when it involves your own people who thought they were doing a good job for the customer, it was nothing short of a disaster. It showed all of us that you must *check, check, check* even when it involves a customer.

A punishing lesson

As a result of this, and at the insistence of the U.S. Air Force legal department, we had to agree to suspend or relocate some very fine employees. This ended up hurting some outstanding people who were just trying to do a job for a key customer. There is no question in my mind that this could have all been avoided if our technical representative had told us what he knew two years earlier, before the scandal broke. He claims to have tried, but I never heard from him or even knew of his existence until well after the whole mess unfolded. He certainly had several opportunities to speak his piece, as he returned to the United States every year for an in-depth discussion of his job. He also signed annual government compliance certificates certifying that he was not aware of any wrongdoing. Apparently, taking advantage of U.S. whistle-blower laws for personal gain was a far greater incentive than telling the truth. As a personal lesson, I found that regardless of how honest and well-intentioned you or your people are, sometimes the enthusiasm for doing a good job for the customer can get in the way of using a reasonable check and balance system.

The net result was that a number of excellent people were hurt in all of this, not because they played a part in the whole mess, but because it happened on their watch. Just like the leader of an air force squadron has to take his lumps when the people in his unit do something wrong, the U.S. Air Force wanted us to do the same here. I only wish that Dotan and Steindler would have been punished more severely (Dotan received a 13-year sentence) and that we could have gone after our technical representative for concealing evidence that would have prevented the whole mess from escalating the way it did.

I do not think I was any more devastated in my career than when I had to be the one who informed some of my key people of their fate. For once in my life, I really regretted that I was close to these men, as it was very difficult to take an action against them that I really did not believe in. Within a short time, we lost the management of the best military engine business in the world. I know I will never completely get over this incident. I am sure the people who suffered job actions will not either.

Sometimes things like this happen that cause you to question everything that you are doing. I had basically relied on trust to guide my philosophy of leadership. Just as I had been treated by my leaders, I let people pretty much run their own shows, believing that we were all working together with essentially the same values and objectives. The Dotan affair was like our Vietnam. I think it caused everyone at Aircraft Engines to look at the way we did business and say, "Maybe we're not as good as we thought we were." I was so distraught over letting the

Overcoming a Failure of Integrity

company down that I was ready to resign, but I felt the business needed me to stand up and lead us back on track.

As a result of this, we did create and extensively publicize a hotline that made it possible for people to tell an independent party of any possible indiscretions without fear of reprisals. We also became the epicenter of a company-wide effort to place integrity at the forefront of every leader's business practices. Integrity and compliance—the spirit and the letter of the law—still take top billing each time the leaders of Aircraft Engines lay out business plans to their subordinates. This was a lesson a business never wants to learn twice.

I was helped considerably during this period by Bill Conaty, my HR leader, Ted Boehm, the chief counsel for Aircraft Engines (later replaced by Henry Hubschman when Ted returned home to Indiana), and Ben Hineman, GE's corporate counsel. This case was a tough pill for all of us to swallow, but it showed the great strength of GE in the instant learning that we mustered. I can only hope that this experience and the procedures we put in place as a result will prevent any similar problem in the future.

Chapter 12
Going Down to Go Up

Staff reductions can occur under a variety of names—re-engineering, cutbacks, rightsizing, downsizing, etc. They can also happen for a variety of reasons. Some examples might be when new technology comes along and allows a business to do the same amount of work with fewer people, when orders do not come in as fast as you planned, or when the direction of a program is changed or the program is completed or terminated. Each of these plus many more possibilities demands a reduction in manpower, both in the factory and in the offices, and this is a prospect that is not easy to deal with for either management or the people affected. While leaders would like to ensure they keep the best people even if they shuffle them around, that is not always possible. Good people may not want to be shuffled. Managers to whom they could be transferred might prefer to keep the team they already have. In addition, the best people can usually get other jobs with relative ease, and so they are not likely to live in limbo.

Many feel that hiring people is a much easier prospect. Unfortunately, while that may be true most of the time, people selection skills are hard to come by. I was fortunate to take a course on intensive interviewing from a man with the unfortunate name of Dick Fear. While this course gave me an excellent guide for selecting people, it was not a panacea for everything. For one thing, in today's legalistic environment it is hard to trust referrals because the people for whom the person previously worked can seldom be relied on to give a true evaluation. All of this makes me believe in recruiting the best people straight from college and putting them through a company training program to see how they react under a variety of circumstances. Unfortunately, this process makes downsizing, which is already difficult, even more emotionally charged, as these people feel such a loyalty to the company that they would be completely let down and demoralized if they were let go.

At GE, we always tried to let go the bottom 5% of the performers first. People go through many phases in their lives, and it is important that there is a constant evaluation going on to ensure that the people and the jobs are compatible, and so we did this on a regular basis even when we weren't downsizing. Fairness is an absolute essential. Personality quirks should never be a cause for removal. Sometimes this is hard to avoid, however. It is especially onerous if one of the people working for you is doing a good job, but your boss feels that he is not and should be let go. I have found that as much as you want to help and protect this person from the boss, you are all better off helping him find another job. As difficult as all of this sounds, it is important to remember that to succeed we must deal effectively with a variety of personalities, especially with the personality of the boss.

Downsizing

Taking all of this into account makes downsizing very difficult for everyone. Still, you have to try to keep the best people, as the whole purpose of downsizing is to protect the total business entity. It is also important to note that proceeding slowly so that the downsizing is stretched out in waves does no one any favors, even though some people get to keep their jobs a little bit longer. In the end, the lower-ranked people whom you let go first get the good outside jobs that the more proficient people who are let go later would probably have gotten had they been in the running. For the business, the bottom line has to be that the people you have left can and will protect and grow the enterprise. That is all that matters.

I took over GE Aircraft Engines after the reign of Gerhard Neumann at the end of 1979. (Actually, I took over from Fred MacFee following his one-year interregnum.) Neumann was a legend not only for his business acumen at GE, but he was also known around the world for his adventures prior to coming to GE. Gerhard had built an organization that suited his militaristic background; however, in spite of his being very frugal, the organization had much more middle management than I thought was necessary. As a result, when I took over, I felt a need to restructure the management system. This meant reducing the executive-level management from 550 down to approximately 440. This not only had a big direct cost impact, but we also found there were fewer meetings, fewer trips, and much lower costs than just the reduction in the salaries of these people. For the sake of the business, I had to make these reductions, but at the same time, I had to convince the organization that our new team would both protect the company and grow the business. (We, of course, did exactly that.) Many of the executive class saw the light and moved on. The people on Neumann's staff who felt they deserved to be his replacement more than I did were also helped to find other employment, both inside and outside the company. These people had been part of a very close team and had delivered a great deal of success at GE. They proved their value by continuing to be successful at GE or for other businesses.

The transition from the Neumann era to my administration was facilitated by Fred MacFee, who was an interim boss while Neumann was recovering from the effects of his heart valve replacement. Fred unfortunately still had Gerhard looking over his shoulder and never felt completely free to do what he wanted to do. That changed when I took over. At that time several key people retired, giving me the opportunity to put my own people in some key jobs. As my HR guru, Don Lester helped me do this fairly painlessly, although we did lose a few key players to Pratt & Whitney. While they were top-notch engineers, these people simply did not fit into plans to grow our team. I wanted a team that really worked together and was able to have productive discussions without a lot of "I told you so" coming from the ranks.

The bottom falls out

Team building and downsizing are hard to do at the same time. We were really put to the test in this regard in 1989 and 1990, just as we were feeling our oats after getting a lot of good commercial engine orders. Our military business had also been doing well. Then we experienced a couple of hits that required us to turn around fast. The first was the end of the Cold War, which totally torpedoed our military sales. This was followed by the Gulf War, which raised fuel prices to such an extent that the airlines were operating unprofitably. Lockerbie and other terrorist incidents convinced many potential passengers that air travel was unsafe.

The double whammy of reduced margins and declining sales drove our airline customers into a financial death spiral. Wherever they could, they canceled or delayed orders for airplanes bought in anticipation of rosier times. Engine sales, linked to aircraft deliveries, declined dramatically. Our sales revenues plunged.

This presented us with a serious money squeeze. We needed to fund development programs (primarily the GE90 and CFM56 derivatives) and keep our operating income commitments to the corporation. On top of this, we had grown our organization to position us for growth, and we now had just too many people. On the positive side, our Continuous Improvement efforts were showing us how to deliver more products and services with fewer resources. By starting our Total Quality thrust during the preceding good times, we gave ourselves the option of pursuing an expanding market, had that occurred, without adding resources or, conversely, downsizing without seriously hurting our core business in a market decline.

A philosophical conflict

Total Quality, or the Continuous Improvement Program, as taught by Dr. Edwards Deming, was something of a populist movement that melded the aspirations of labor and management. Labor, in Deming's view, would only be committed if they trusted that increased efficiency would generate expanded opportunity, not fewer jobs. Deming and his cohorts pointed to Japan's alleged policy of jobs-for-life as the successful evolution of Total Quality. (No thought was given to the ranks of Japan's minimally rewarded sub-tier suppliers who lived a precarious existence dependent on the winds of economic change and the vagaries of the karetsu mentality.) Our Total Quality consultants were Deming adherents, and they vehemently opposed any reduction in the labor force. (The disagreement was so bad that we terminated what had been a relatively long-term relationship over our decision to continue with our downsizing plan.) Many of the people they had helped train, largely those in the ranks of hourly employees and salaried individual contributors, also embraced Deming's notion of the sacredness of employment. My staff and I had always said, as loudly and as clearly as we could, that we were in a cyclical business and that continued full employment would never be a guarantee. People hear what they want to, however. The mere thought of massive downsizing was considered a breach of a tacit promise by many.

In truth, we had built a good team and, while some players were clearly better than others, we did not really want to lose anyone. There seemed no way to resize our workforce without hurting people. Try as we might to be fair, some people were going to end up hating GE and the leaders of Aircraft Engines. As we began the process of downsizing, a cartoon was making its way through the informal communication channels. It showed a father standing at the head of the dinner table with his wife and children sitting by their empty plates. "I've had some bad financial news," the father is saying, "and I'm afraid I have to let one of you go." At one point, a scrawled cardboard sign hung in one of the many empty office cubicles. It read, "Will Audit for Food." Clearly, some of our former employees saw us as casting out members of the family to beg for a livelihood. In the face of these implications, the ones who remained took little joy in their survival.

My staff in early 1990: front, left to right, Frank Pickering, Lee Kapor, me, George Krall, and Ed Bavaria; rear, left to right, Chuck Chadwell, Bob Stiber, Ted Boehm, Brian Brimelow, Bob Turnbull, Tom Cooper, Dennis Little, Bill Vareschi, Frank Robinson, Bob Gerardi, and Bill Clapper.

Many of the most highly qualified workers, believing that they were seeing the handwriting on the wall, decided to get out of the aerospace industry altogether before their turn for termination came. It was difficult to prophesy for whom the bell would toll next, and so we could issue no real promises. We lost many good engineers whom we would end up needing in the years to come. On top of all this, my staff and I were getting ready for retirement.

I had initially planned on retiring in 1991 when I was 60. George Krall, Ed Bavaria, Lee Kapor, and Frank Pickering all retired in relatively short order, and we lost a very key younger man, Bob Turnbull, all in about the same time period. I felt obliged to stay and bring the new young team of Chuck Chadwell, Bob Stiber, Jim Williams, and Mike Lockhardt up to speed. It was tentatively planned that Lockhardt would eventually succeed me, but it soon became obvious that he was not the team builder we had thought he would be. I was encouraged by Jack Welch to stay, and deep down inside, I really didn't want to retire anyway. By the 1992-1993 time period, however, it became obvious that we would have to bring in someone from outside GEAE to run it. Again, I felt responsible to the team I had grown, but I also felt an obligation that the massive downsizing take place on my watch so that my successor, who turned out to be Gene Murphy, could build his own team without having to fight the downsizing problem.

So we downsized, and like a natural disaster, downsizing consumed everyone's thoughts for a while. During that time there was probably much more work done in the ranks brushing up resumes and talking to old friends than on engines. In the end, however, we not only weathered the storm but we also came through it with a team that was more than capable of taking full advantage of the upturn that inevitably followed the downturn. We actually dropped from more than 40,000 employees to just over 20,000 in a one-and-a-half-year period.

Making money

It is also true that, while our airline customers were losing more money in the course of three bad years than commercial aviation had made in its entire history, we were turning a profit. I think this caused no small amount of resentment, as if, because the industry was bleeding, we should bleed in sympathy.

There is a bit of sports doggerel that says it is not important to note whether you win or lose but rather how you play the game. From my experience as an amateur athlete, I would say that both are important, but being known as a good sport in defeat is certainly cold comfort. When you become a professional, however, winning is everything. It is your job. In business the score is kept in dollars, and the leadership of GE Aircraft Engines is professional. We were neither inhumane nor unethical when it came to making some painfully difficult business decisions, but we were professional. I also believe in a saying that Neumann was fond of quoting, "If you show me a good loser, all I see is a loser."

When you are downsizing, it is important to make sure your base technologies are protected. We were helped in this regard in that we were in big development programs on the GE90 for the 777, CFM56-5 for the A340, CF6-80E for the A330, and the CF34 for the regional jets. Funding all of this development became very hard on the rest of the organization, however. One innovation that helped keep this technology growth alive, in spite of the fact that it was to be used over fewer programs, was our Centers of Excellence, which I have already mentioned. I was worried that, as the more experienced people left through early retirement, we would not have enough expertise to spread out under our current program-based organization. By tying design engineering, manufacturing, and materials closer together, the Centers of Excellence helped keep our technology base intact. Of course, when you do something like this, it puts an added strain on the project and systems managers because they have less formal control. It also requires that you let some key individuals who have the potential to become a systems or project manager rotate through assignments in the centers.

Stepping into the light

As we emerged from the downturn and were growing our organization to cover expansion, it became obvious that we did not want to expand our two big plants in Evendale and Lynn. We had seen with Rutland, our blade and vane plant in Vermont, just how much more productive a smaller operation with between 1200 and 1800 workers could be compared to the big plants—usually 20–30% more productive. One reason for this I believe is that in a smaller plant the boss of the plant can really get to know all of the people on the team and work with them to grow stronger. We also found that the union officials in the big plants did not want to cooperate with us to improve quality and efficiency in our Workout and Continuous Improvement programs.

In addition to the blade factory in Rutland, we had a factory making small parts in Hooksett, New Hampshire, a turbine blade factory in Madisonville, Kentucky and a casings plant in Wilmington, North Carolina. This type of decentralized structure helped us decide where to downsize. We used productivity as our first requirement, which meant that more of the downsizing went on in our bigger plants. I wish I could say that this handwriting-on-the-wall made Lynn and Evendale more productive, but while there was some improvement, there was always a long way to go to catch up to the energetic satellite plants. Ultimately, many of the assembly operations, which had long been the mainstays of Lynn and Evendale, were assigned to smaller plants. A facility in Durham, North Carolina was created to assemble GE90s, and our Strother, Kansas service facility began assembling F110s and CF34s.

One of the corollary events accompanying the startup of the GE90 program was our exponential expansion into the service sector of the business. We were always in the service end of the business to some degree, but it became clear that repair shops were not only a good source for our parts and component repair sales but were also a way to offset the fact that we were selling engines at well below the margins necessary to keep GE profitable. By getting more fully involved with the total life-cycle operation of the engines, we could more readily afford to take less profit from the initial sale of an engine if we knew we stood a chance to recover that margin over the life of the engine. Of course we were greatly helped in developing state-of-the-art proprietary repair processes by the fact we were already producing very reliable engines and had background knowledge on their strengths and weaknesses.

Our approach to vigorous entry in the service arena was to acquire a huge number of existing independent engine service shops and then transform them into the GE mold while retaining the good things that made them successful before we acquired them. Following any vast expansion program there is, of course, consolidation. In spite of expanding business opportunities, there are invariably some people who will no longer be needed after the expansion stops. Luckily for me, dealing with that downsizing is now someone else's problem.

Lessons

Letting people go is always difficult, but no matter what you do, you owe it to the business to maintain your key technology base. We were able to do that with our Centers of Excellence together with our chief engineer's office. At GE, we were also helped by the fact that many of the people we had made available to other parts of the company returned to us with an expanded perspective when an opening presented itself, many to fill key positions, such as George Krall, Ted Torbeck, and George Oliver.

Another valuable lesson was that an orderly procedure for layoffs helps sustain the morale of those remaining. We accomplished this by setting up a relocation center where employees could get help with resumes and make contacts with prospective employers. We even had recruiters from other businesses and from other parts of GE come in and conduct interviews. None of this was very pleasant for the people involved, but it is interesting to note that many of them found better jobs than they had had at Aircraft Engines due to good GE training. Many who were simply stagnating in their jobs were able to use the impetus of downsizing to spring into a more fulfilling career.

Going Down to Go Up

An invaluable asset in the process of helping decide who should go and who should stay was Bill Conaty, my human resources manager. His input to the managers who had to make the actual decisions kept as many of the personality issues out of the equation as possible.

While downsizing convinced me that we needed to keep our key-technology people, it also pointed out that many jobs could be handled by subcontractors at substantially lower costs. This was especially true in the financial and shop services parts of our business. We also found that some of the lower-technology parts of our engines could be designed by other lower-cost companies.

Downsizing also showed communication and training to be an ongoing need through the process. Extensive communication was essential to rebuild the remaining team, and training, especially in Six Sigma techniques, was absolutely essential to reduce the impact of the inherited workload.

Downsizing gave me no joy, but after it was over, I felt very proud of the remaining members of the team. They went on to grow the business without adding large numbers of people. The subsequent expansion of the services sector of the business initially added lots of people. The leaders at GE are now working to reduce much of that hands-on labor by focusing on the things we are really good at—parts design, system support, and management.

CHAPTER 13
The Lessons and Legacy of the GE90

In 1988, when we first pulled a team together under Bruce Gordon to study an engine full of potential use for a new airplane, we really did not know what new airplane we were talking about. We were working closely with Boeing, McDonnell Douglas, and Airbus to predict the future needs. It soon seemed obvious that the efficiency of twin-engined airplanes flying long-distance routes over water would be extended to ever-bigger airplanes. Airbus did not want to see us follow that approach since such an airplane would compete with their four-engined A340, which they had just launched. They saw our efforts as something like consorting with the enemy, forgetting that we designed and built the CFM56-5C for that aircraft. We even offered to make a smaller-diameter version of the engine we were considering for Boeing to make Airbus's A330 more competitive with Boeing's new airplane. Boeing, naturally enough, thought creating a wide-bodied, twin-engine airplane was a great idea, and they strongly encouraged us.

We responded by studying different derivatives of the CF6 to see how we could make that engine both bigger and better. Each study hit a stone wall at about 75,000 lb of thrust. The barrier was the size of the hole in the center of the engine through which the shaft from the low-pressure turbine to the fan passed. The hole was simply not big enough for a beefy shaft to carry the enormous torque to drive a low-speed, high-thrust fan.

When we had originally designed the CF6, the hole through the center was intended for a fairly long shaft that would transmit torque to the fan of an engine producing 40,000 lb of thrust. As the engine power requirements go up, the torque on the shaft goes up. Developing continually stronger material helps, but even so, the shaft twists between 13 and 20 deg when the engine is running. In addition to twist, the shaft has little eccentricities or harmonic resonances, and it will whirl away from its axis at places. This is controlled with bearings, but the frames to hold the bearings add weight and the bearings add friction, and so you only design in what you need. As the torque demands on the shaft go up, the tendency to whirl goes up. Putting in another bearing—and the frame to hold it—is essentially redesigning the engine.

We could have kept the same diameter shaft, run it at higher revolutions per minute and lower torque, then reduced the rpm at the fan with a gearbox, but this would add weight and complexity. One of the beauties of the jet engine is that it really does not have parts that rub on one another other than the bearings. Gears are being rubbed all of the time, and as a result, gearboxes are usually problems waiting to happen. I did not want to go in that direction as that would add a weak link in what was otherwise a simple and highly reliable design. We did not want to buy that kind of trouble especially when the first application would be on a twin-engine aircraft.

Because this hole was part of the basic architecture of the engine, it meant that there was no way to grow the CF6 if we wanted the fan to be directly driven by the low-pressure turbine short of adding a beefier shaft and changing the bearings, the frames, the cooling system, and the turbine disks—in effect, a new engine.

Another complication was that the CF6 was already fairly long. If we added additional compression stages to get the 20:1 pressure ratio in the gas generator that we felt we needed to make the engine deliver the required thrust, it would get even longer. In short, the growth potential of the engine beyond 75,000 lb of thrust was limited.

We were confident that long-range wide-bodied twins would enter service with engines at about 70,000 lb of thrust to begin with, but we were also certain that the requirement would expand to 100,000 lb of thrust very quickly and then more slowly escalate to about 125,000 lb of thrust. Growing the CF6 as far as possible would get us just barely over the threshold of the minimum entry point. Clearly, a whole new design was in order if we wanted to compete to power more than the initial wave of these new airplanes.

A technological challenge

The way would not be easy. We knew that to run at the engine temperatures we foresaw, we would need to use a new high-temperature alloy, but we felt that was doable. Besides thrust, another major requirement was that the engine be quiet. Thrust and noise are interconnected. To get more thrust, the engine must produce a gas flow that is hotter and under greater pressure. In a high-bypass engine, that energy is then transferred to the fan, which moves high volumes of relatively low-speed cool air to push the airplane. One way to convert the increased energy of the gas flow into more thrust is to make the fan bigger. All other things being equal, however, the bigger the fan, the faster the tip speed of the fan blades. Most engine noise is caused by the tips of the fan blades as they break the sound barrier. Designers can mollify that noise by putting in a long duct full of sound baffling for the bypass airflow. Another way is to slow the fan down.

If the fan is turning relatively slowly, it has to be really big to move the amount of air necessary to deliver that 100,000 lb of thrust. Again, all things being equal, bigger fans mean heavier fan blades. We had had experience with lightweight, composite fan blades on our unducted fan demonstrator, and felt we could successfully apply that technology to this new engine.

With the basic design for this engine in hand, we again went calling on the airframe manufacturers. Airbus still wanted nothing to do with it. Boeing, on the other hand, was now looking at a number of configurations, and one of them was destined to become the big twin-engined 777. In 1990, we announced our intent to design and build this new engine, the GE90. From that point, we began working closely with Boeing, defining our engine as they defined the 777. After many meetings, I believe I convinced Boeing that they needed to bite the bullet and go ahead with the airplane. I got the impression that Boeing would have liked to work with us exclusively on the powerplant from the beginning, but it was obvious that the other engine manufacturers wanted to be involved.

The Lessons and Legacy of the GE90

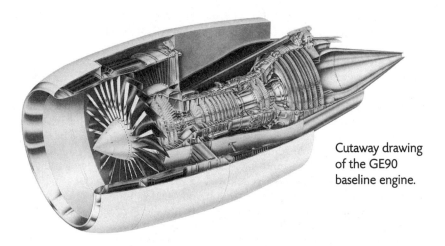

Cutaway drawing of the GE90 baseline engine.

Pratt & Whitney and Rolls-Royce, however, were neither in a position nor of a mind to commit the requisite $1.5 billion or so to design a brand new engine. They each decided to make a virtue of necessity by offering a derivative growth engine using relatively tried and true technology that they promised offered no surprises. In truth, their existing engines had nowhere near enough growth potential to approach the 100,000-lb range without major modification.

Boeing's initial offering of the 777 was to be a good aircraft for relatively short intercontinental city pairs. The thrust requirement for this first version would be about 70,000 lb, well within what we could have developed with a derivative of the CF6—and well within what Pratt & Whitney and Rolls-Royce could deliver with their derivative engines, as well. Because of the GE90's more advanced technology overall, we expected to have a fuel burn and quietness advantage over the Pratt & Whitney and Rolls-Royce engines, but we were also creating a platform that was capable of growing to well over 100,000 lb. As a result, our engine was going to be much heavier than the competitors' engines. We also were starting with a blank sheet of paper, and it was easy for our competitors to paint bleak pictures of missed deliveries as we sorted out the new technology our new engine would require.

The first competitions

It soon became painfully apparent that our preliminary design and business plan would do us no good unless we found an airline willing to be a launch customer. United Airlines was the first 777 customer and the first competitive battleground for the engines to power it. We worked very hard with the technical people at United, but United is a traditional Pratt & Whitney customer. (In fact, Boeing, United, and Pratt & Whitney were all part of the same company until the U.S. government broke them up sometime in the 1930s. So, the roots go deep.) Pratt & Whitney had a lot of engines on United's airplanes, although we did have some CF6s on DC-10s with United and some CFM56s on some of their 737s.

United had us, Pratt & Whitney and Rolls-Royce cooped up in three tiny rooms in their Chicago head office, and the United negotiating team bounced from room to room. It was one of the most belittling and demeaning competitions our team, headed by Ed Bavaria, had ever participated in, not just for us but for our competitors as well. The price competition was fierce, but in the end Pratt & Whitney came up with a good plan to integrate the engines for United's 777s with an existing 747-400 engine deal. At the time of this competition, United was about to exercise their option on some 747s they had previously agreed to buy with Pratt & Whitney engines. The combination deal that United and Pratt & Whitney concocted was something we could not match, and by the end of negotiations, we were merely foils—and uncomfortable foils at that. Pratt & Whitney got the sale. This was a blow for us, because we felt we had the best engine, but it also pointed out the new buying techniques that we could expect from some of the airlines.

In spite of this setback, we had a lot of design work going on, and we were beginning to order material to start on the development engines. Our relationship with Boeing was outstanding, and they really wanted to see us on this airplane. So, we started looking toward the next big buyers for the 777—All Nippon Airways (ANA) and Japan Airlines (JAL). Both of these companies had strong relationships with Pratt & Whitney, although we had made several recent engine sales to both companies. The Japanese tend to be conservative, however, and betting on a new engine would be a stretch for them. In addition, as with United, they only wanted a short-range airplane, putting our heavier engine at a disadvantage.

We also failed to see that ANA and JAL were planning to cooperate and order the same airplane–engine combination. We thought there were two competitions when there was actually only one. Pratt & Whitney apparently saw what we did not see. They brought all of their marketing effort to bear on the first competition, ANA, where, based on customer relationships, we should have stood our best chance. Pratt & Whitney got the sale to ANA—and by default, the sale to JAL as well.

British Airways

We had lost at United and in Japan. The next competition was at British Airways (BA), generally an exclusive Rolls-Royce customer. We absolutely needed a win or we were really out of business on the 777, and this must-win in Rolls-Royce territory was not the best of situations to get in. To our advantage, Rolls-Royce had grown complacent about winning at BA. They were treating this sale as if it were a done deal. A side issue also came into play on our side. GE was looking to expand its Engine Services operation at the time and was interested in purchasing BA's engine overhaul facility in Wales. These were entirely separate issues at the beginning of negotiations, but they were destined to come together just as Pratt & Whitney sold engines for 777s and 747s together at United.

Over the years, with the help of James Barrett, our man in England, I had spent a lot of time developing a good relationship with Lord John King and Colin Marshall, BA's leadership. I had also worked closely with Boeing to help sell them some CFM56-powered 737s. In addition, I had tried to make sure that we built relationships at all levels as we competed for a follow-up buy of 747s they were making in 1988. Our main objective was to show the people at British Airways and

The Lessons and Legacy of the GE90

in the British government that we really did "bring good things to life" with our engines. In the end, although I think we got close, we lost the 747 order. I like to think politics played a part, but I was never sure.

In any event, we were not going to let politics be the deciding factor on the GE90. On the one hand, GE was building up its employment in the United Kingdom. On the other hand, we had learned that BA was looking to divest itself of its engine maintenance work just at the time that GE Aircraft Engines was looking for a chance to expand our service business in that part of the world. Even with these factors on our side, many of my colleagues felt I was crazy for trying to penetrate this Rolls-Royce stronghold.

I was encouraged to believe that a win was possible when Lord King and Colin Marshall introduced us to some important ministry people. It also helped that Rolls-Royce engines were not doing that well on British Airways airplanes. Rolls-Royce was neither fixing the problems nor keeping BA informed of what was going on. They had forgotten the basic lessons of *take no customer for granted and treat each customer as Number One*.

The competition for the 777 at British Airways was reaching its final stages in the summer of 1991. After I had visited with all concerned in late July, I went off to Cape Cod on vacation, where I kept up a continual dialogue with all of the principals by phone. In fact, one day I was on the phone for more than three hours. I got such a cramp in my hand that I had a speaker phone installed.

It soon became apparent that we were reaching the end of the competition, and a GE team set off to complete the final negotiation. Since the drama was drawing to a close, I thought it best to be in the thick of things. I arrived in London at about 11:00 in the morning, but in the middle of the negotiations, all of the BA representatives disappeared, leaving us to wonder what was going on. Apparently, Rolls-Royce had gotten the idea that we were winning, and they wanted to make a last-minute bid to counter what they thought was our proposal. Perhaps Rolls-Royce was asking British Airways to close their eyes and think of England. (Apparently the prime minister had even gotten involved.) When the BA people went to visit the Rolls-Royce negotiators, we knew we needed to do something dramatic to stop the wavering.

Bill Vareschi, our finance manager at the time, was with the negotiating team. I told him to see if he could get a check for a down payment on the Wales overhaul facility immediately. It was now about 1:00 in the afternoon. At about 5:30, Vareschi came back with a big smile on his face and a very large check in his hand.

By 11:30 that evening, we were still waiting. Finally, Colin Marshall, BA's president, called us in. "This is really a tough decision for us to make," he began. "We like your engine, we like your people, and we don't like being considered locked-in customers by Rolls-Royce, but we're still a little bit shaky on this thing."

About then John King, BA's CEO, called on the phone. "How is it going, Brian?" he asked.

"Fine," I said. "I think we're close, but, John, I just don't understand you guys. This seems like a pretty straightforward decision. Do you have a political problem with buying our engines?" No, he said, that was not an issue.

"Well, I've got a problem," I continued, fingering the infamous check. "I've got this large check as a down payment on the Wales facility. What the hell am I going to do with it?"

"You have what?" King asked.

"I've got this check. If you guys don't buy, I've got to take it back."

"Well, okay, Brian." King responded, ending our conversation. With that, I left the room.

About five minutes later, Colin emerged and said, "OK. It's a deal." (Actually, our guys had been snooping around looking for something to drink, and they found the celebratory champagne being prepared, and so we had already begun to breathe a little easier before Colin Marshall's real announcement.) We signed on the spot and turned over the check around midnight, leaving British Airways to tell the prime minister in the morning. (Their chief financial officer was left to sleep with the check!)

GE and Rolls-Royce

Rolls-Royce was actually pretty good about our victory on their home turf. I think they realized that if they win at American Airlines or United, we do not raise a commotion. If they can compete freely in the United States, we should be able to compete freely in the United Kingdom. Over the years, it had been rumored—I think largely by people at Rolls-Royce—that we at GE were trying to drive them out of business. It is true that we have always sought to be aggressive competitors, and just like a sports team that is looking for a perfect season, we always plan to win each and every competition. But just as with sports, the game would not be much without opposition. Usually, we are able to keep our competitive urges on a business basis, but sometimes they do seem to get a little personal.

Much of the fabled ill will between GE and Rolls-Royce at the time of the British Airways signing stemmed from a joint project we had begun in the early 1980s. At the time, we admired their technology, and our people got on well together. As it became obvious with engine competition on the Boeing 777, there simply was not enough profit margin for three manufacturers to each invest $1.5–2 billion to power each new airplane program, only to give the airlines the privilege of choosing between three engine options. Rather than compete across-the-board in this fashion, we offered Rolls-Royce a revenue-sharing partnership on two comparable engines. They would share in approximately 25% of our fairly new CF6-80C—and receive 25% of the revenue—and we would do the same for their RB211-535. We worked this deal with a very fine gentleman, Sir William Duncan, in 1984. Duncan had worked in the United States and had a good feel for international cooperation. Unfortunately, he died suddenly, and Francis Tombs took over. Tombs was more influenced by the people in the home office at Rolls-Royce.

The cooperation was all going fairly well until we saw a sales brochure in which Rolls-Royce planned to offer a derivative engine, which was going to cost them a lot of development money, to compete with the CF6-80C they were supposed to be supporting. In their brochure, they rated the CF6-80C as a fourth-place engine in the competition. There was no legal restriction to keep them from doing this, but it certainly seemed like a breach of good faith to us. I confronted the Rolls-Royce leaders over this, but when I showed them the brochure, they said, "We've never seen that before," which we knew to be a fabrication.

In retrospect, I guess I reacted a little like a wronged lover. "Look," I said. "Either we're going to work together, or we're not going to work together. Let us

The Lessons and Legacy of the GE90   133

know what you want to do." The Rolls-Royce leaders were happy to have the opening. They opted immediately for divorce. This put us in a difficult position because Rolls-Royce was our sole source of supply for a number of parts. We had to do some expensive scrambling to get another source geared up.

Cutting our losses, we decided that we were not going to make a big deal out of the collapse of our failed marriage, but at the Paris Air Show that year, I was asked about it, and I let my heart do the speaking rather than my head. I was quoted in the newspapers as saying, "The deal had made a lot of sense, and we thought we were let down." At the end of the statement, I added the impolitic quip, "But, we don't get mad, we get even." This was interpreted as hurling a gauntlet, which was not the intent. I should have said, "We'll see who wins in the marketplace."

In truth, I think Rolls-Royce would have benefited more from the cooperation with GE than we would have. In any event, we reverted to the three-way engine competition for many sales, and the pricing war continues as the standard way of doing business in the aircraft engine industry. As a result, even if a new engine is successful, it usually takes 15–20 years to break even. That is a long time to be paying out of your pocket until you start to see a profit. In any event, Bill Duncan and I believed that, by working together, we could make a really great engine in the future.

Back to the GE90. Once British Airways said yes to the GE90, we were about to embark on that lengthy business process again. We had made the initial sale. Now, all we had to do was deliver the goods and hope to make some more sales.

With drawings in hand, we laid out a schedule and began finalizing arrangements with our partners. Snecma, our 50/50 partner on the CFM56, was to be a major player on the GE90. They wanted a larger part of the engine, but ultimately we settled on 25%. Fiat-Avio, the Italian aerospace company, received a smaller share, as did Ishikawajima Heavy Industries (IHI) in Japan. Motoren-und Turbinen-Union (MTU) of Germany also had a share in the low-pressure turbine. The relationship with the partners was destined to be a tough one in the early going. Nobody could have foreseen the levels of price concessions we would have to make to sell in a brutally competitive market with three contenders. In addition, while we thought we were designing to cost specifications, it turned out that we would have a pretty expensive engine unless we did some major cost reduction.

MTU slips away

We were just getting started with our detail design when we were suddenly faced with the prospect of losing MTU as a partner. We received word from our representatives in Germany that United Technologies Corporation (UTC), Pratt & Whitney's mother company, was trying to cut a big deal with Daimler Aerospace (DASA), MTU's holding company. Our relationship was really quite good with the MTU people, but we were never really that close with the upper management of DASA. We were to learn once again that you have to cover all of the bases in relationships with partners.

I flew over to see the leader of MTU, Herr Dunkler, a fine gentleman who, unfortunately, did not have full knowledge of what was going on between UTC and DASA. Dunkler gave me his assurance that MTU was still on track to be working with us on the GE90. A week later, however, he called to tell me that there was to be a joint cooperation between UTC and DASA. MTU would now be working closely with Pratt & Whitney. I told him that we were very disappointed and that we would have to consider a suit against them as they had been fully integrated into our GE90 team. We did not want this technology going to Pratt & Whitney. In light of this development, we also would have to discontinue our arrangement of having MTU produce turbine blades and vanes for our CF6 engines, since we did not want them to benefit from this program only to feed earnings to their work on Pratt & Whitney engines. This would be a big blow to the production side of MTU's business, as more than 50% of their production work was CF6 parts. This prospect really shook the MTU team, who wanted to work with us, but their bosses in Stuttgart were adamant that they work with Pratt & Whitney.

All of this resulted in MTU and ourselves having a big negotiation a couple of days later at the Lynn plant to see how they could save their production line, with us being compensated for the problems they had caused us. After several hours of hard negotiating, we agreed to let them continue their CF6-related production—but at a much-reduced profit margin. While this did not solve any GE90 problems, it did improve our margins on the new CF6 engines and spare parts. The loss of MTU as a GE90 partner cost us precious time as we now had to get our other partners to do some of the low-pressure turbine parts. Ultimately, we were able to get the GE90 back on schedule with our other partners, but it took some scrambling.

While our negotiation with MTU was a tough one, we maintained a good relationship with the people there through it all. On the other side of the issue, the relationship between MTU and Pratt & Whitney never really developed as anticipated, and so while we lost a partner, we never lost a friend.

Funding crisis

At about the same time as the MTU crisis, other things totally out of our control turned around to bite us. As I previously referred to, the collapse of the Soviet Union spelled the end of the Cold War. Our military engine business went from $4 billion a year to half that. Then the Gulf War started. The price of fuel went through the ceiling. As if to add insult to injury, many thousands of would-be passengers, afraid of terrorism, stopped flying. Our airline customers were losing money faster than they could shovel it out the door. The last thing they could afford was a new airplane. Airlines were pushing out and even canceling orders we thought were like money in the bank. All of the money we had counted on to develop the GE90 had dried up. This put tremendous pressure on the whole organization and forced us into a cost-reduction mode. It also made us look at all of our programs and delay or defer some of them.

While we were striving to maintain relationships on the GE90 team, we were also working hard to keep up relations with Airbus. We had been developing the CF6-80E1 for the Airbus A330, a program that started off with great potential when we signed the original deal, but began to pale due to the way Airbus was marketing the aircraft. We would have preferred to switch Airbus to a refanned GE90 for the A330, which would have improved it, especially from the growth potential, but Airbus was having no part of it. They felt that course would be an endorsement of the GE90 on the 777, which in turn would hurt their four-engined A340. Slowing down work on the CF6-80E1 only served to confirm their resentment over our relationship with Boeing on the GE90.

More technological challenges

Good airplanes grow, and they require good engines to grow with them. Unfortunately, a lot of the potential customers did not see it this way. That is why we were very thankful to British Airways for their far-sighted perspective that enabled the GE90 to really get started. They provided some outstanding people for our *Working Together* team who were as dedicated as we were to making the program a success.

At this time, the leader of the GE90 program was Ron Welsch. Ron Welsch and Ambrose Hauser put together a good program and were the key people involved in integrating with the Boeing 777 effort. With help from many people throughout the organization, they came up with a credible engine that was recognized as the technology leader with the growth potential to take it well into the 21st century.

The gas generator, or core, of the engine was scaled from the E^3 (Energy Efficient Engine), a demonstrator program that had been headed up by Marty Hemsworth and partially funded by NASA. We had great confidence in this compressor that produced 20:1 pressure ratio with nine stages of blades, still a very advanced concept even though more than 15 years had passed since that demonstrator program was started.

We began with a layout of an engine that would meet the original requirements with 76,000 lb of thrust, all the while keeping in mind that we wanted the ability in the essential architecture of the engine to grow the thrust to 125,000 lb. With a preliminary design in hand, we ran simulations to analyze various stresses and the aerodynamic flow through the engine components. As part of a design review, we identified critical technologies that we had to perfect before we would feel confident enough to put them together as a complete engine for testing. One of the components we did that with was the compressor.

Testing a compressor as a freestanding component is quite a task. In the actual engine, the highly compressed air coming out of the compressor has fuel added and is ignited to deliver a high-speed gas flow that spins a turbine that powers the compressor. In a component test, there is usually no combustor or turbine—just a compressor. To spin a compressor at operational speeds takes enormous power. In Lynn, we would test our small engine compressors by driving them with a steam turbine that came from a navy destroyer. This compressor was too big for that, and so we had to use its own combustion system and turbine. Logically, you want to run the compressor test before you run the complete engine test, but in actuality, you are making and assembling both at the same time, and so there is no room to stop and start over if you want to meet your committed timeline.

Testing is something like alchemy instead of pure science. Sensors have to be added inside the engine to measure stress, strain, temperature and aerodynamic flow. Just putting sensors on the airfoils is a fine art. They do not last that long in operation, and so it helps if you have someone who can quickly understand the test

GE90 compressor being readied for testing.

output and come to a conclusion about what adjustments need to be made before the sensors are destroyed.

The sensors were not a problem on the GE90 compressor test. The test was going fine—then the test rig broke because it had not been designed to be strong enough. The failure destroyed the compressor. We had to huddle and consider whether we had enough data from the partial compressor results to commit to full core engine testing. We decided to go ahead. When we ran the complete engine, we discovered some seal leaks that we would have discovered earlier, and so we did have some fixing to do, but by going ahead with the engine test we kept the development process largely on schedule—something that was essential for customer confidence.

It takes a lot of creative work to build a new engine, and, of course, usually you like to have all of the bugs worked out of the most unique new technologies before applying them to the design. Sometimes, however, you end up developing the technology as you are designing the part. That is the way it worked on the composite fan blade for the GE90. (The technology was not exactly new. We had tried our first composite blade on the Quiet Engine Program, but in that case we were not required to do foreign object damage testing, and neither did we have to do it on the UDF—the Unducted Fan engine—as that fan was considered a propeller. So the technology was not fully matured.) It was important that the fan blade be made from carbon-fiber composite material because of its high fatigue strength and lower weight. We also calculated that composite fan blades would be more durable in the long run in terms of resistance to damage from things such as nuts and bolts on the runway and birds flying in the air. All of this turned out to be true, but getting there was a lot harder and took a lot more time out of our schedule than we really had to give.

The development of the composite fan blade was a remarkable piece of work by both our design and manufacturing people. The composite material is essentially strands of carbon fiber in a resin matrix that has been formed into sheets. The sheets are layered over one another in such a way that the criss-crossed fibers will give enormous strength while they transmit the loading on the blade to the dovetail where it is attached to the hub. After the layers of composite material are layered to form the blade, the whole thing is heated and put under pressure to merge the resin and lock the carbon fibers in place. The result is a strong blade that is much lighter than titanium, the lightest metallic alternative. As an additional advantage, a graphite composite blade does not fatigue, because it is not a crystalline solid. If a titanium blade is chipped or nicked, that chip or nick can easily propagate into a crack. With a composite blade, the nick does not propagate through the fibers and no crack forms. (The resin matrix is softer than metal, however, and so we did add a thin titanium strip along the leading edge of the blade to protect it.)

Besides carrying the loading to push air and power the airplane, blades must also be able to withstand being struck by objects the airplane might encounter. The most severe of these threats is birds. A bird hitting the tip of a fan blade at flight speed is a tremendous impact. It starts an oscillation in the blade that puts an enormous stress on the base where it attaches to the fan disk or hub. Our first bird impact and foreign object damage tests in September 1993 were not as successful as we had expected they would be. They showed that we needed to reinforce the blades in several areas. In response, we did some excellent modeling on the computer that helped us put the reinforcement in just the right places. To make these blades, we set up a factory in San Marcos, Texas, where we increased the yield and got the cost down. We then felt confident we could make composite blades at lower cost and with greater reliability than a hollow titanium blade, the previous standard.

The composite blade has proven to be a very important technology breakthrough for this type of high-bypass engine, much to the credit of several very creative engineers. It is also important to note that, regardless of how realistic we think our tests are, all of our fans seem to do much better in service than their simulated tests results would predict.

After correcting our problems, we put our core engine back on test on 16 March 1993. This was a crucial part of the program because, while we were testing components, we were building the full engine for its first test in parallel. It was important that we have a successful core test before we could run the full engine, and we did that just 13 days after the core test. This took a lot of effort on the part of the test and evaluation people as well as the manufacturing team, who had to produce many parts that were constantly being updated as they were being produced.

The first run of an engine is always a very dramatic moment for any of us that have ever been involved in an engine program. We not only had to ensure that all of the parts were made, assembled, and instrumented correctly, but we also had to have test people who understood the limits of the parts they were testing. The first test of an engine is a real test of the team you have working on it. Of course, management and marketing people are taking shallow breaths and crossing their fingers when the startup takes place as well.

We ran the GE90 for the first time on 29 March at our outdoor test facility in Peebles, Ohio, and by 2 April we had achieved a record-making 105,400 lb of thrust. The whole team drew great satisfaction from this as well as breathed a collective sigh of relief. We had passed the first step toward integrating what we wanted to achieve in this new technology engine. We could now point at a real engine and say it did what we said it would. The third-generation engine designers and builders who put the GE90 together were also blessed by excellent design reviews by people who played a big part in getting us into the industry in the first place—Marty Hemsworth, Gene Stoeckly, Joe Alford, and Mel Bobo from the 1940s and 1950s, and Jim Tucker, Lee Kapor, Tom Donohue, and Roy Smith from the 1960s through the early 1980s. It is important to have people who did this work in the past participate in design reviews, as it is never possible to put down on paper, or in a computer, all of the things that go into the design of great engines.

Our euphoria was soon replaced by a big letdown when we had two major problems in July and August 1993, which took some fast action by everyone to fix. Again, the use of the chief engineer's office and our "old experts" proved invaluable. After we made some changes in how we assembled the engine and modified

GE90 composite fan blade.

The Lessons and Legacy of the GE90

The first GE90 in testing at Peebles, Ohio.

a couple of designs, we were able to run the first engine in the new test cell built for the GE90 in Villaroche, France, by the Snecma team. As usual their test went very well and really made them feel a much bigger part of the program than just a parts supplier. We also ran the first GE90 at the IHI facility in Tokyo on 16 May 1994. Again, the teamwork among the Japanese engineers was up to the task, and as with the French, it was very gratifying to all of us to see that IHI felt themselves to be a completely involved part of the program.

At this point, we were into the certification program. FAA certification involves more than merely running the engine. It must be proven over a range of temperature, speed, and environmental conditions to be certified. In addition, the first application for the GE90 was to be the Boeing 777, a twin-engined airplane. Aircraft with two engines are required to fly routes that are close to airports should an emergency occur that shuts down one engine. After an engine had proven its reliability, the certifying agencies can grant it an extended twin operation (ETOP) rating of 120 or 180 min. A 180-min ETOP rating allows routes three hours away from the nearest emergency airfield—enough for trans-Atlantic flights. Instead of accumulating in-service flight hours toward this rating, we committed to enter service at 180 ETOP. To do this, we had to fly the engine and fly it a lot.

The first step in this direction happened on 6 December 1993. This was the first flight of the GE90 on our modified Boeing 747 flying test bed. This is an airplane we leased from our sister company, GE Capital Aircraft Services (GECAS), that had three standard engines—actually, Pratt & Whitney engines—and a fourth pylon to which we could attach our test engine with full instrumentation.

Soon after we got our first GE90 into the air, the main drive shaft failed. Controlled panic followed. It turned out that the shaft was just not getting enough cooling air. We modified the flow into the shaft and put another engine in the air. All went well.

As the engine starts accumulating cycles, people start looking for parts that show unusual wear. The engineers might say, "We're getting too much rub on the tip of this blade," or "I've got to increase the cooling air to that shroud." We might have some vibration to trim out, or the flow in the combustor might need some tweaking. All the while this is going on, the engine is logging flight time and accumulating data.

The flying test bed had a family tie for me in that my son, David, negotiated the deal for GECAS with Ron Welsch, who was the project manager at the time. Of course, David was as thrilled as the rest of us when the first flight was successful. The flight-test team at Mojave deserves special credit as they worked with Boeing to pull together a very meaningful flight-test protocol. The aircraft gave us the opportunity to fly the engine throughout the entire flight envelope and let us do a real evaluation of all of the engine's flight characteristics much better than tests on a fixed test bed would have. We were able to check engine re-light schedules, control schedules and stresses in various parts of the engine under all operating conditions and g-force loads, as well as do the environmental testing. Most important, of course, was the fact that we could simulate landing in crosswinds and the aircraft's rotational effects on the GE90 before installing the engine on a new 777. As with any test protocol of this type, it is the teamwork of everyone from the designers to the flight-test crew that can make or break a program. This is also a spotlight for the engine in its first appearance "in public" as we invited the flight-test people from Boeing to put our new engine through its paces.

The GE90 on our B747 flying test bed.

The Lessons and Legacy of the GE90

While flight testing is going on, the engine is also undergoing testing on the ground. This is the point at which we are firing 3- and 5-lb birds at various places on the fan of a running engine. (We had done this test on a rig with spinning fan blades earlier and made modifications to the blades. It is important to do as much testing on components as possible before risking a $20 million engine.)

Engine tests are often very dramatic, and no matter how much calculation and pretest work is done, the result is never sure until you see the birds fired into the engine and the engine continues to run. Because the GE90 has a large capture area because of its big fan, by FAA guidelines, it had to withstand more bird strikes than engines in past tests—eight instead of six.

We conduct this test on an outdoor test bed with the engine operating at speed facing into huge fans that simulate some of the speed of the aircraft in actual flight. Dead birds are placed in Styrofoam boxes and fired by air cannons directly at the front of the engine. The Styrofoam box breaks away, and the bird goes into the spinning fan. Initially, we thought we could get away with shooting chickens into the engine, but it turns out that chickens do not have the same bone structure as flying birds. We then decided to use sea gulls. That brought flack from the environmentalists, and we had to get special permission to capture sea gulls to complete the federally required tests.

Besides birds, the engine must ingest ice, hail, and rain and keep operating. The engine must also sit in below-freezing temperatures and then start on command. That winter was unusually mild at our Peebles test facility, and we were worried that we would not have a day or two that was cold enough to conduct the tests. Luckily, the weather turned cold for the minimum time we needed.

In addition to bird and foreign object ingestion tests, the FAA requirements had become much more stringent with regard to what is called a blade-out test. In this test, a fan blade is explosively separated at the point where the dovetail at the root of the blade fits into the fan hub. While we did not feel the composite blade would have this type of failure, both the FAA and we felt it was necessary to show that the engine could sustain the separation of a fan blade in flight without allowing any uncontained damage that could potentially affect passengers or the rest of an aircraft's operating systems. The test is very dramatic. It is also very expensive. A fan blade in a rapidly turning engine is blown away at its root where it attaches to the platform. As a special condition, we were allowed to perform the test by separating the blade in its flowpath instead of at its dovetail, as that would be the most vulnerable spot in this incredibly strong composite blade. This test essentially destroys a $20 million engine. Watching this happen first with naked eyes, but more dramatically in slow-motion films, gives one a real understanding of the tremendous expertise involved in making something as dynamic as an aircraft engine hold together. We passed the fan blade-out test on the engine in November 1994, and I was enormously proud as the skill of the engineers who designed these parts was vindicated.

At the same time as we are making adjustments to the engine in the testing program, the people in the technical publications department are determining how to disassemble, maintain, and reassemble the engine and documenting that for us and the customer. And the marketing team is out there trying to sell more engines. Everything is running in parallel. About 18 months before you expect to certify the engine, you have to commit to the number of production engines in the initial run,

GE90 undergoing ice ingestion testing.

as it takes that long to get castings and forgings and to make the fan blades. The partners, like IHI and Snecma, have to be just as ready as GE to move into production on schedule, and so developing an engine requires masterful timing and coordination.

The period from March 1993 until February 1995 was full of great successes (as well as an inconvenient high-pressure turbine blade failure that really taxed everyone involved), and the test program for the GE90 led to passing the FAA certification tests. This is an accomplishment in itself since the requirements are tough in their own right. In addition to this, however, we also did enough cyclic and systems testing in a program that was closely monitored by Boeing and British Airways to ensure we could go into service with a 180-min extended twin-engine operation rating.

The ETOP program was coordinated by George Pirtle, son of John Pirtle, one of our retired senior engineers and a former program manager. We have many cases of multigenerational enthusiasm for aircraft engines in all of our plants. I think it is a good thing, and I can only hope this kind of heritage will persist in the future.

Leading the development program

As I noted, the design of the engine was conceived by a team headed by Ambrose Hauser, but as I have found in the past, it is not always the people who conceive the ideas who are best to put them through the complete program. Preliminary design work takes one kind of engineer—someone who is conceptual

The Lessons and Legacy of the GE90

and more systems oriented. Detail design follows, and it takes another kind of person—someone who is pretty disciplined. This is where the engine gets its final sizing, and so the ability to coordinate with the airframe customer is vitally important. The detail design team must also be willing to do battle with Manufacturing and Test and get on with the job of delivering a preliminary product. Bruce Gordon headed the detail design team. From this assignment, Bruce went to Lynn to head our Small Commercial Engine business, and Dennis Little became the first official GE90 Project Manager. The project manager oversees all aspects of the engine's life cycle, from engineering to procurement to manufacturing to marketing to cost accounting, without having most of the people doing these jobs reporting to him.

In August 1992, Dennis became the vice president of our marine and industrial engine business, and Russ Sparks became the GE90 project manager. In April 1993, Mike Benzakein became Russ's engineering manager. Mike had an excellent record of putting various versions of the CFM56 through their development programs, and he was highly regarded by Snecma and IHI.

Also in September 1993, I was made Chairman of Aircraft Engines to help transfer the reins of the business to Gene Murphy. Gene was new to aircraft engines, but he had had extensive experience in GE Aerospace. Being chairman instead of CEO freed me to follow the test and development program of GE90 a lot more closely. While I had not done the design work, my direct connection did give me a great deal of satisfaction as I watched the engine develop. I was also able to run some cover for the team with Gene Murphy and Jack Welch, who sometimes could not understand why designers were not perfect. There are more than 34,000 different pieces in a GE90, and each of them has to be designed to work perfectly all of the time. While Six Sigma can dramatically improve a product's performance, there is still a lot of work in maturing an engine design that is hard to put a number on—and without quantification Six Sigma is somewhat blind.

First flight

For me, the culmination of the engine program was the first flight of a GE90-powered B777. This took place 2 February 1995, the day after I retired from GE. It was not only a great finish to my career, but it was also the first first flight of an aircraft powered by a GE engine that I ever witnessed in person, and I was hoping I would not jinx it. During all of the previous first flights I was always too busy (or so I felt) to attend personally.

Alan Mulally and his Boeing team did a splendid job in coordinating and designing this superb aircraft, and while Pratt & Whitney did fly before us, I felt this was the first real flight of Boeing's 777. I was lucky to be asked later in 1995 to fly to London's Heathrow Airport to show off our GE90-powered 777. The airplane got a great reception from the British Airways people, and it was most rewarding to have Lord King and Colin Marshall present to meet the aircraft. (It was also interesting that, when we arrived, an old 1931 Austin was there. This was the same car as the first one I owned.)

The airplane was full of test equipment, and this gave my wife, who accompanied me on the trip, a real appreciation of what it takes to make an airplane and

First flight of a GE90-powered Boeing 777.

its engines work together. I felt for the first time that she really understood why I have such a great love for this industry and the technology and people involved in it.

Going into service

The job does not end when the engine goes into service. In fact, we figure that about 60% of the engineering expenses occur in the development phase, and 40% come after the engine is in service. The first issue we faced going into production was cost. The people involved in the detail design and development phases were supposed to be keeping an eye on cost, but the pressure to meet the schedule seemed a lot more important. As a result, we had an outstanding engine that cost too much to make, and we had to spend a lot of engineering effort getting the cost down in the manufacturing end of the equation.

For the first few years after an engine goes into service, we put technical representatives every place that airplanes with this engine are going to land. Their job is to check the condition of the engines, ensure that any problems are fixed and report back on the general effectiveness of the product support operation. They sometimes uncovered problems that involved the human interface with our product—things such as mechanics putting the cap on the oil tank improperly. We had to do some redesign so that the cap could not be put on incorrectly or forgotten altogether. The same was true of inspection plugs.

There were also genuine design issues that we had to address after the engine was in revenue service. We found that the variable stator bushings in the compressor were in need of replacement after only 10,000 hours of operation, which used to be a lifetime for us. Now, 20,000 hours is a minimal expectation. We redesigned the stator bushings and got their life up to everyone's expectations.

After engines have been in service for a long time, we do an analytical teardown to let our engineers see what parts would benefit from redesign before the next release of the engine. In that way, we continually make the engine more reliable and more easily maintainable. This, of course, is a mixed blessing, as we make a significant part of our income over the life cycle of an engine on replacement parts. And the engines just seem to run forever.

The legacy

The first GE90-powered 777 entered service in November 1995 and has had an excellent record ever since. I was disappointed that American Airlines and Delta selected the Rolls-Royce engine for their airplanes, since American, along with United, launched us back into the commercial engine business in 1968. Our CF6 engine served them well then and continues to do an outstanding job to this day. I suppose the Rolls-Royce folks felt equally jubilant about getting them as customers for their Trent engine as we did when we won at British Airways. I still feel that, as Ed Bavaria says, "There are some airlines that are happy to have our engines and some that wish they had them!"

Since I retired from GE, Gene Murphy moved to a job in corporate management and subsequently retired. Jim McNerney followed him as CEO of Aircraft Engines, and upon Jeff Immelt's accession to the chairmanship of the GE Company, Jim moved on to head 3M and was succeeded by Dave Calhoun.

In the years after my retirement, the GE Aircraft Engines leadership and the Commercial Engine Operation under Chuck Chadwell continued to do a great job of growing the GE90. It is now up to the thrust size that really makes it the best engine for the 777. As a result, Boeing selected the GE90-115B to power their new, larger 777-200LR and 777-300ER aircraft exclusively. It was very gratifying when Jack Welch called me after he closed the deal with Boeing to tell me we had won this exclusive arrangement. I felt vindicated that I had made the right choice in the size and configuration of the engine. The new, bigger 777s have been bought by JAL and ANA—who switched to our engine—as well as Air France, EVA Airways, and Pakistani International Airlines. For me, this is very satisfying. These successes go a long way toward making sure that we have not only a technical success with the program, but that we also make it a financial success.

The GE90 is the first engine brought out by the new team of young engineers who are continuing the tradition of GE Aircraft Engines. I am sure they will do a great job in continuing to grow the business. Russ Sparks and Mike Benzakein, as a result of their good work on the GE90, have moved on to higher-level positions, but the team now running the program has made a seamless transition with customers and the design team. While the GE90 is often looked at as my engine, it is clear that it was created and supported by many other people. Key among them were Jack Welch and the corporate team, who, despite many heated discussions, really were with us all the way.

A view toward the future

The GE90 is the state-of-the-art commercial production engine. It exists because of a huge commitment of money and talent—a commitment that continues as the 115,000-lb thrust version approaches certification and revenue service. The higher thrust GE90 is emerging at perhaps the worst time in commercial aviation history, a time of massive financial uncertainty after which only the fittest and most adaptable of airlines are likely to survive. This is a hard time to sell airplanes and engines, no matter how reliable and economical they may be. It seems like a good time to rest. This, however, is not what is happening.

In addition to the GE90-115B, Aircraft Engines is hard at work developing the GP7000, a joint project with Pratt & Whitney, for the enormous Airbus A380; the CF34-10 for the Embraer 190 and 195 and a Chinese regional jet, the ARJ21; and, with Rolls-Royce, the F136 for the Joint Strike Fighter. And after these engines are flight tested and certified, there will still be more work to do. As I look toward the future, it is clear that the demand for greater fuel efficiency will be both perennial and attainable, although the marginal return for the engineering effort will be ever smaller. Another avenue for improving the efficiency of commercial engines is reducing the nacelle size for a given airflow to improve the aerodynamics of the whole engine system. Nacelles are currently thick because acoustic treatments and protective shielding are thick. As advancements in materials technology allow those materials to get slimmer, nacelles can be thinner, and more of the fan will be useful in moving air.

Then there is the issue of durability. The current crop of engines stay on-wing—that is, do not need to be removed for overhaul—for sometimes as long as 30,000 hours. As that becomes the standard, the bar is sure to rise. To reach those greater expectations, we must both reduce the number of parts in the engine and improve them. By reducing the number of parts, we can also make engines cost less to build. As I noted, the competition for new engine sales is so intense that engines are often sold with little or no margin for the seller. Engine makers traditionally counted on years of spare parts sales to make back their development costs and ultimately show a profit. As engine parts last longer, *ultimately* goes farther and farther into the future. Because any qualified shop could repair any manufacturer's parts—although not necessarily with the same part life or performance—the only defense for the engine maker is to ensure that parts that wear out, or the best processes for repairing them, are patentable to protect the investment in technology. This does not always endear us to our customers, who would like low costs at both ends of the deal.

Making significant engine improvements will not be easy. A present-day compressor may be running at about 86% efficiency. That means that 14% of the air entering the compressor is lost through leaks or other demands on the system such as cabin air and engine cooling air. To improve that efficiency by 2% to 88% sounds incremental, however, to reduce the losses from 14% to 12% is a 15% improvement—which would be no small accomplishment. Such a compressor would present a higher mass of air to the combustor. Because the mass would be greater, the velocity of the gases could be lessened and the same force would still be presented to turn the turbines. Lower velocity means lower temperatures. Lower temperatures mean lessened cooling requirements. Conversely, more air flow at the same velocity offers the potential of higher thrust.

As one part of the engineering team works on improving compressor efficiency, another is working on the profile of the gases coming out of the combustor to eliminate hot spots on the turbine blades. Others are working to improve engine stability to minimize axial flex and whirling, usually by making the engine shorter. At the same time, someone is looking at the balance sheet. Can we make it that much better without putting too much money in? Improving basic designs is a difficult chore, but if you can deliver a 10% improvement in thrust, reliability, or fuel efficiency at the same price as your competitor, customers will buy your engine. Even 8% or 5% might do it.

The legacy of the GE90, or of any engine project, is that improvement never ends. The need to strive for just a little bit more is the curse and the joy of the engineer's heart.

CHAPTER 14
Politics and the Engineer

Many people enter the field of engineering because it has a certain mathematical purity about it. They see themselves as focused on optimizing the performance of shape, structure, and material in absolute ways and are pleased because what they do can be directly and unequivocally evaluated. If the criteria for a design are clearly stated, the true engineer would never accept that the decision as to which design best met the criteria was a matter of opinion. The truth of politics, on the other hand, is seldom a matter of black-and-white hard facts. While political truth can be analyzed effectively many years after events, the rightness of today's political decisions often seems the stuff of opinion. The give-and-take of politics is generally foreign to the engineering mind. Engineering ultimately leads to manufacturing in most cases, and manufacturing leads to selling. When what you are selling represents a huge capital investment and is a matter of national or regional pride both for the seller and the buyer, the power and uncertainty of politics are not far behind. So it is in the high-technology, huge-expenditure world of jet engines.

When I began my career in engineering as an apprentice, I had no thoughts other than to work on airplanes and engines—and play sports. Even though I saw the start of World War II as a boy in school, the war was never explained, and I had no real interest in the politics behind it. We simply knew that, under the leadership of Winston Churchill, who gave many inspirational speeches that had all of us young people convinced, and with a lot of help from the Americans, we could never lose the war. We really were not told much (or, if we were, it was not in the sports pages).

Just after the end of the war, there was a general election in the United Kingdom, and Clement Atlee and the Labour Party beat Winston Churchill and the Tories. I was curious and tried to understand why or how this could happen. It became apparent that many of the men returning from the war were upset by their standard of living, especially the miners and the manual workers. They also felt the Tory party should have done more to help us be better prepared than we were for the war. This brought up the whole discussion of who should run what. For some reason these workers felt they would have a better life under a party that believed the main mining and transportation companies should be nationalized. They thought they would then have some say in how these companies were run. While this may have worked as a short-term fix, I am certain it did not help in the long run. I feel sure that nationalization did get rid of some awful management in some of these companies, but the bureaucracies it built up were even worse. These industries formed boards in London, and soon the workers were more remote from their bosses than they ever were before. There is no question, however, that the standard of living for many of these workers improved.

It was not until I came to America, and my wife and I were invited to a dinner with a lot of other employees and their wives, that I got a sense of how much more dialogue there was in the American management system than I had seen at home in England. I developed a feeling that participation at all levels in both management and politics was both possible and encouraged. I may have been lucky that the company that employed me in the United States was GE, as I have subsequently seen other companies that are not as democratic. You may ask what all of this has to do with politics. Well, it is my belief that in the United States you can, if you work at it, achieve anything you want with your life. The opportunities are limitless, even today, whether it is in engineering, finance, or politics.

Clement Atlee

As a result of my move to the United States, in early 1960 I had an opportunity to meet with Clement Atlee, the man who was my anti-hero, displacing Winston Churchill after the war as prime minister. This meeting came about when I was invited to a meeting at the British consul's house, where I was asked to keep Mr. Atlee company prior to his giving a speech in Cincinnati. I found him to be much quieter and less outspoken than I had seen him in the United Kingdom. He obviously was still in favor of nationalization and felt it was best in the long run for the people. I told him I disagreed, and he said everyone is entitled to his or her own opinion. That was my first opportunity to really meet a politician, and I was not that impressed. Over the next several years I was exposed to many American politicians and their aides, and I must admit that I felt many of the aides were better informed than their congressmen and senators.

Lobbying for the sale

During a campaign to keep the B-1 bomber in production, we visited a congressman from a southern state who put his feet up on his desk, showing us the holes in his shoes and his one brown and one blue sock. He did not seem interested in what we had to say, even though the project would have generated work in his state, and he told us to meet with an aide who was his expert on defense. We expected to meet with some crusty old guy, but instead the aide was a young lady who couldn't have been older than 22. We gave her our pitch, and much to our surprise, the congressman voted in favor of the program. The lesson here was never to underestimate the power of the aides.

Of course, during the Great Engine War a whole cadre of GE executives made many visits to Washington to discuss with various committees why the United States should buy our engine for the next batch of F-16 fighters. We received a lot of attention from everyone, as they didn't want to make another mistake after the problems they had had with the Pratt & Whitney F100 engine. Our presentations were well received except by the representative of the state where Pratt & Whitney builds their engines. We had excellent coordination between Harry LeVine's team in our Washington office and the Secretary of the Air Force, and there were some hearings where our people did a great job against their Pratt & Whitney counterparts. Pratt & Whitney resorted to a lot of whining comments about GE having all of the business. They said they would only have two legs on their three-legged stool if GE got the contract. (This was after many years and many millions of dollars spent by the air force trying to get Pratt & Whitney to fix their engine.) As a result of our good work at politicking and Pratt & Whitney's failings, the government representatives believed they needed a change. From the engineering standpoint, the most important thing we did was run a series of tests with our demonstrator engine. We also completed a flight-test program paid for by the military that showed our engine was not only generally the most desirable but also capable of operational maneuvers Pratt & Whitney's F100 could not match. In addition, we invited congressional representatives to a series of plant visits both to Evendale and to Peebles, our outside test facility, to show how we had proven the strength and reliability of our engine. The engineering, marketing, and lobbying teams did a great job, and after about three years of compromise, we won the major share of the next fighter engine buy. This had GE not only selected to power the F-16 but also chosen to re-engine the F-14. (We had also run a very successful series of tests on the F-15 but to this day have not been able to get the U.S. Air Force to switch. If that had happened, I am sure GE could also have gotten a number of export orders for this aircraft powered by our engines. We did get one order from South Korea in 2002.)

Politicians have a lot on their plates and cannot be experts in everything they have to vote on. This is where the aides come in. However, I did get a feeling during the Great Engine War that the politicians who really made it their duty to understand the pros and cons were the ones that eventually swung the votes. There is no question that our communication efforts in this regard were better than Pratt & Whitney's.

International intrigues

Several other political associations come into play when you're selling, especially internationally. People often feel that the airlines always have their own choice of aircraft and engines, but in the end politics usually comes into play. One campaign that was particularly upsetting to us involved Air New Zealand (ANZ). We had a great relationship with Air New Zealand that began with the delivery of their DC-10-30s and grew from there. After a few years, Morrie Davis—who, when we first started, was not the boss of ANZ but was vice chairman or CEO—eventually became chairman. He and Neil Burgess had a particularly strong relationship. More importantly, our CF6-50 engine had done a really good job in New Zealand on their DC-10-30s. We had also supported them in their shop and really showed them that, even though New Zealand is about as far away from Cincinnati as you can get, we would give them full support. We entered a competition to provide ANZ with engines for some new 747s, and based on our relationship, we felt we had an excellent chance of winning. At a Conquistadores meeting, Morrie told me ANZ's board had decided in our favor, and they just had to get clearance from Prime Minister Muldoon. Halfway through the meeting we heard that Mrs. Thatcher had called Muldoon to offer that, for the United Kingdom's promise of buying lamb and cheese, he should buy Rolls-Royce engines for his 747s. Muldoon agreed, much to the disgust of the ANZ guys who were never really happy with the Rolls-Royce engines they ended up with. We subsequently sold them some engines for 747s, but the earlier missed sale was very important to us as we were just getting a foothold on the 747. I guess we were naïve in that we didn't think there was any political agenda. If we had, we would have at least tried to make some representation to the government of New Zealand. It turns out that in dealing with sales in foreign countries, you can never really expect that there are no politics involved.

In the 1980s, after the Great Engine War, there was a competition in Israel in which Israel was going to build a new fighter, and they had selected the F404 for their airplane. (Wiseman was Secretary of Defense at the time.) We thought we were doing very well in the competition. Then we heard that Morrie Zipkin, who had left GE and gone to work for Pratt & Whitney, had come up with an engine derivative of their F100 called the F120. Morrie had made several visits to Israel and was a great friend of General Dotan. Pratt & Whitney was obviously making a lot of progress. My guys asked me if I would go to Israel and talk to the defense minister and shore up what we thought was our winning proposal. When I got there I found out the defense minister had been replaced, and now the prime minister, Menachem Begin, was also minister of defense. I received an audience with Mr. Begin and told him about all of the great attributes of our engine. I said, "You know this is really not a tough decision for you, Mr. Prime Minister." He was very congenial. His English was not bad, and he said, "No, Mr. Rowe, this is not a tough decision." So I said, "Do you think it was as tough as the one when you decided to blow up the King David Hotel?" which was the headquarters of the British prior to Israel's settlement. "No, that wasn't a tough decision," he replied. "That was an easy one." I thought, oh well, we'll have to give this a real try. We promised them everything, but unfortunately—or fortunately it turned out in the end—we did not win that order. Pratt & Whitney won the competition and spent a lot of money on the program, but nothing ever materialized. Instead the Israelis bought more F-16s, a lot of them with our engines. So, politics had a way of rearing its ugly head, though I did have an opportunity to meet Mr. Begin, who, while he was a small man, was very impressive and interesting to talk to.

In selling commercial engines, we only occasionally thought that we would get embroiled in as many political situations as we actually did. One I will never forget involved Airbus Industrie. Airbus was trying really hard to sell airplanes in Iran, and at that time there were U.S. sanctions against Iran. The United States did not want American companies to sell anything there, and so we were asked by Airbus to see what leverage we could bring to bear because the Iranians wanted our CF6 engines on their airplanes—Airbus A300s. I went to the White House and spoke to the assistant undersecretary of state. We talked at length about how we should be able to sell our engines on the Iranian airplanes because if they were not sold with GE engines, they would be sold with Rolls-Royce engines. It didn't seem to me there was any political advantage one way or another except GE and the United States would lose some potential sales. After all, we were selling the engines to Airbus, not to the Iranians. Anyway, to cut a long story short, we won that one. About a month after that, Boeing had an Iranian order for 20 737-300s, which seemed to me pretty innocuous. I considered that we ought to be able to convince the White House again that we should be allowed to sell airplanes over there. Again, the argument went, if we did not sell them, Airbus would probably sell A320s with the IAE engine, even though they had some Pratt & Whitney content in them. (Because of the small Pratt & Whitney content, I don't think the United States could have controlled much of that sale.) Anyway, we met with

Brent Scowcroft and his National Security staff and many others in the White House. I went with Dean Thornton and the Boeing representative, and we had a terrific meeting. We thought we had solved the problem, but after we left apparently the politics of the situation prevailed, and someone felt our engines and airplanes could be used against us. We did not get government approval. Even though it may appear that I was only worried about selling our engines and not the political situations, that is not so. If there is an embargo or sanctions against a nation, unless you get the whole world to support the sanction, all you do is rob your own people of an opportunity to sell and develop relationships. I firmly believe that, when people have your equipment, you build relationships—and that that is much better for the total political situation. But my feelings may not always be right in that regard.

Habibie and Indonesia

In the early 1980s, we were asked by the Indonesian minister of technology, Dr. Rudi Habibie, if we could put our engines on their CN235 aircraft and help them in coproduction of the engine. At that time Indonesia didn't seem like they would be in the running for building an airplane. On my first visit there to the Industri Pesawat Terbang Nusantara (IPTN) company, about 3 years prior to Habibie's proposal, it was just a small company with a hangar where they were repairing some old airplanes. It was amazing to me that Dr. Habibie had the ambition to build up an aircraft industry there as he had many other industries. He had spent a lot of time in Europe. He had been educated in Germany and really had excellent relations with the Germans. He also developed good relationships with Boeing. He even worked out an agreement under which he would share the building of the CN235 with Construcciones Aeronáuticas SAC (CASA) of Spain.

After a lot of hard selling, we beat out the Allied Signal engine, and we got our engine on the CN235. We helped set up the shop where they would do the final assembly of engines, and that part of the relationship went very well. In the meantime, I had become personally enchanted with Dr. Habibie because he was this small man with big, big plans for his country. Even though he was at times belittled by many, he did build a tremendous technical infrastructure for his country, not only in aircraft but also in many other technologies. I think this base is something the Indonesians will be building on for many years to come. Politically I never thought Habibie was that astute, but he had a very close tie with Suharto, the president of the country at that time. Suharto took Rudi, his mother, and his family under his wing after his father, who was a general, was killed. As a result, he had the ear of General Suharto, who really gave him a pretty free reign on running the aircraft industry as well as many of the industries he was involved in. Most importantly, Habibie educated a lot of people who became obligated to him and treated him like a god. Rudi has a very neat wife, Ainun, who was very approachable. She was a pediatrician and, most importantly, kept Habibie on his toes from a health standpoint.

After we got to know him and had started assembling engines, we found the airplane they built, the CN235, was quite a nice airplane. It was probably a little bit underpowered, and we didn't have a growth version of the engine. That eventually put another engine on it, although we continue to sell engines on this aircraft. We also helped them on the flight-test program, letting them use one of our flight-test people to do the first flights of the airplane, which showed there was improvement needed in the flight control system. Without improvement, it was just too hard to handle the airplane. Over the course of several years, we built up a very strong relationship, and as a result of that, we eventually sold engines to Garuda-Indonesia. I'm absolutely convinced that Habibie was a big help to us there. He really enjoyed the relationship with GE and also built a strong association with Jack Welch and many others at Fairfield, even though they sometimes felt that Habibie was overly ambitious for his country. But people forget that unless you have a go at something, you never know if you're good, bad, or indifferent, and I think he gave it a real swing.

Eventually Habibie decided he would build a turboprop at IPTN. He had received orders for work from Boeing and Airbus in his factories. He also did final assembly of some German helicopters, the CN235, and the CASA 212 airplane. The organization was growing, and they went from almost zero employment to about 17,000 people working at the plant. After I retired, I was asked to help him set up a facility to build an airplane called the N250 in the United States from components they would send over to us. In all honesty, I could never understand why he would want to do that. It just seemed a very political thing to do because there is no question that the airplane could have been produced much cheaper in Indonesia, ferried over and finished in the United States. They certainly did not need to do the final assembly there. The people in Mobile, Alabama, were very keen for him to build a factory there, and so they made some pretty generous concessions. This gave me an opportunity to work again with Jack Edwards, who was one of the strong supporters of GE in the great engine war.

Our plans were going along fine, but it was obvious the airplane was slipping. I arranged for a team of people under Lou Harrington, a former Douglas man, Linwood Lewis, a former GE man, Russ Heil of Delta, and Ken Knutsen, a former Boeing/Rohr man. They formed the nucleus of the team, which sat down with the Indonesians to come up with a business plan and to get some semblance of discipline in the organization. They had done a lot of good design work, and the first airplane was built. Unfortunately, the discipline of making sure the parts of the airplane matched the parts as defined by the drawing was never really enforced. No matter how much the American part of the team tried to do this, it took a long time to convince the Indonesian side that it was much easier to do the job in an orderly fashion. The thing we found out was that Habibie was one of the few Indonesians who knew how to say yes and no. The people working for him were not that strong and waited for him to tell them to do everything. As a result, just as in any organization, unless people have delegated authority, it is really difficult to get a complicated program going—and building an airplane is a complicated program. The cost of the program was not significant at that time. However, there were quite a lot of comments by anti-Habibie elements about wasting money on an airplane when they should have been doing things for the poor people. While I do not disagree with that, those same people conveniently forgot that the many countrymen he was educating would eventually end up as the backbone of Indonesian society once they got their political mess out of the way.

Also during this period, IPTN sold many of these airplanes to Garuda, and we were starting to build a good relationship with them. We had semi-annual meetings with Habibie on the project, and, of course, I had Harrington, Heil, Lewis, and Knutsen spending a lot of time in Indonesia helping them. The program was stretching out, however, and the money was being questioned very vigorously, but Habibie continued to have the support of Suharto. As time went on, it became clear that we should not build a facility in the United States, and so we really dragged our feet as we tried to get Habibie to see that it was not the right thing to do. In the end it was lucky for both Mobile, Alabama, and Habibie that U.S. production fell through, because the business was just not ready to build this airplane on any sort of schedule.

They actually started to assemble a third airplane after the first two had flown. The design had some problems, but they were not bad aircraft. Unfortunately, they are still building that third airplane about four years after we stopped the project.

In the meantime, Habibie had built up relationships in the various parts of Indonesia. Many people do not understand that Indonesia is the fifth largest country in the world from a population standpoint. It also has 17,000 islands, many of which are still populated by cannibals. Habibie used to really relish some of the political trips he made as Indonesia's minister of science and technology because he would meet people who looked like they came from the stone age. It was just amazing to see him trying to bring some education to a lot of people who really were not ready for it. They were just simple people enjoying their lives. To make a long story short—most of it is history now—when Suharto was deposed as president of Indonesia, they asked Habibie to step in. Again, Habibie got a lot of criticism from many people around the world claiming that he was a technocrat and therefore not really able to run the country. But to give credit where it's due, in the short period he served as president, he definitely tried to bring in democracy. He did not let the Suharto family have the influence a lot of people thought he would. As a result of bringing in democracy, Habibie was kicked out in the first election because he was associated with the previous government. I think when the final history of Indonesia is written, however, Habibie will be recognized as one of the fathers of democracy in Indonesia, a man who had the courage to start trying to install a democratic government even though he suffered politically as a result.

We forget sometimes that it took England about 2000 years to become what it is; it has taken the United States 500 years, and places like Indonesia and Africa are expected to do it all in 30 or 40 years. That is just not the way it happens. People have to be educated. They must start at the bottom and work their way up. There must be a lot of dedication to eliminate corruption. No matter how primitive societies are, there always seems to be corruption. While Habibie made a valiant first attempt to root it out, it is obvious today that he was not that successful. He did bring democracy to Indonesia, however; and now the people have to find someone who can lead them through their current situation. Of course, that may take splitting up Indonesia. There are a tremendous number of different languages there, and 17,000 islands, and so it's not going to be an easy task to pull them all together. Many people just do not want the sort of technological life many of us in the western civilization have decided is the right thing for us.

Habibie was one of the more interesting people I met outside of GE. He was very knowledgeable on many different subjects. I'll never forget the time we took him to the Woods Hole Institute while he was staying at our home on Cape Cod. The people at Woods Hole put on a great show. He started asking questions, and they were quite amazed at his knowledge. He talked about the El Niño effect, and I think at the end they were listening more to him than he was to them. This is not unusual because any conversation you have with Habibie is 90% Rudi and 10% you—if you're lucky. He was a very enjoyable person to work with and help, and I believe we did help because we certainly helped him avoid spending a ton of money in the United States when he didn't have to. We also saved Alabama a lot of money, although I'm not sure they knew what was really going on.

The inscrutable East

Japan was also a very interesting lesson in how the political bureaucracy really helps industry. Bureaucracy and industry can work very closely together. I think the Japanese do this better than anybody, though to an outsider it seems to add a lot of time to any discussion. Every time we wanted a co-production program with the Japanese, or to sell engines to JAL or ANA, we always had to ensure we made visits to the bureaucrats. We had some excellent people working for us in Japan—Ken Shibuya and many other Japanese as well as Americans—who got us in to meet with these people. Sandy McCord and Wayne Andrews did an excellent job in Japan. We used George Gregor quite a lot in our earlier sales work there.

I did not think too much about all of this until we were trying to sell GE-engined DC-10s in Japan. Mitsui had bought some DC-10s on speculation, and we thought these would go to ANA eventually. As a result, we considered we had an inside track. The Lockheed L-1011 was the competing aircraft for the ANA-JAL sale. It soon became apparent that Lockheed had done a much better job with the politicians than McDonnell Douglas, and so a McDonnell Douglas–GE team flew to Japan for many meetings. We had a Japanese representative there who felt certain that under-the-counter payments were being made. Of course we hadn't heard about it. (I am often very naïve in these matters and frankly feel very lucky I did not hear about some of these goings on that did not involve GE directly.) In any event, there was one dinner meeting where Mr. Tanaka's bagman (Tanaka was the prime minister at that time) was supposed to meet with Douglas's Dave Lewis and Jack McGowen. I don't know whether Lewis and McGowen thought better of it or what, but that meeting didn't take place. As a result, we never did get the contract, which was probably the best thing that could have happened. In the end there was a big scandal about the contract that Lockheed got. By the way, the airplanes that we thought Mitsui bought for ANA were eventually leased or sold to someone else.

Sometimes people who try to use influence, instead of the best product, to make a sale find that it backfires. It really did backfire on the Lockheed people. Even though they sold airplanes in Japan, it hurt several of the key people involved in the sale. In addition, one of the key ANA executives committed suicide as a result of the investigation. We did eventually sell ANA engines for a lot of airplanes—767s, 747s, 747SRs. But every time we tried to sell to the company, we had to ensure we saw all of the bureaucrats and convinced them that we had both the best product and the best interests of the Japanese at heart—and that we would support the products with full commitment. It was a good lesson for us.

We also sold engines to JAL eventually, but it took a history of many tough negotiations with us trying to sell them engines for 767s, 747s, and DC-10s before

Politics and the Engineer

we won their confidence. The JAL management at that time—not the political people but the airline management—were heavily influenced by what they felt was the inadequate job we did for them when JAL bought Convair 880s with our CJ805 engines. We used to make a lot of fence-mending visits, usually with little to show for them. During one competition, Bob Smuland was manager of the CF6 program. We thought we were doing very well in this competition, but the week before Christmas we were told we had not won. The chief of the evaluation team at JAL offered to give Bob a debriefing—on Christmas Day. Bob politely turned him down.

We were always on our toes interacting with the Japanese, even when we were working with them as partners, because as meticulous as they are, it is also very interesting to see how decisions are made there. A lot of people, Deming and others, claim that decisions in Japan are made at the bottom and make their way to the top. I can honestly say that I never saw that happen where we were involved. There was always a lot of discussion at the bottom. The lower-level people took meticulous notes and built a good data file. They met frequently with their bosses who would indicate their likes and dislikes, but in the end, when the bosses, decided what they wanted to do, the lower-level people wrote their analyses accordingly. We caught onto this after a while, and we sent some excellent people to Japan, particularly to JAL, where we eventually got a great order from them. The team was headed by Ed Bavaria and Sandy McCord and a very highly regarded man from Field Engineering, Frank Szecskay. Their efforts ultimately convinced JAL to order 40 Boeing 747-400s with GE engines. I'm sure this really upset Pratt & Whitney because they had long been the traditional engine provider to JAL. Unfortunately for them, they had slipped in giving the support the Japanese expected of them. On our side, we had done an excellent job supporting ANA. (We also built an excellent relationship with Asakura-san, the chief engineer at ANA for several years, who unfortunately died a couple years ago.) While people say ANA and JAL work independently, I'm sure that, to some extent, there was often a splitting of orders to ensure that Japan's major airlines always had their choice of viable vendors for any of these projects.

Our relationship with Ishikawajima Heavy Industries (IHI) on the engine building side was always outstanding. Drs. Inaba and Ossimi were both excellent people to work with, and because they are fervent golfers as well, we had not only excellent working relationships but many interesting golf games all over the world with them.

Our biggest disappointment in dealing in Japan came after Boeing's 777 was launched. Because we now had a great run of successes, we thought we could sell our GE90 engine on that application. Again the Japanese showed us that we could not count our chickens without doing everything to nurture the eggs. The Japanese are traditionally conservative. They like engines that have been proven—and airplanes when it comes to that—but ANA was the launch customer for the 777. When ANA picked the 777 with the Pratt & Whitney engine, we were really disappointed. We were surprised again when JAL picked the same airplane-engine combination since we were coming to expect that the two major airlines would split their orders. We eventually won at JAL with the GE90-115B, and I think we did it on our reputation for quality. I think the Japanese really learned from our experience with the GE90 in service that it is a truly great engine.

Many people will tell you that politics plays little or no role in Japanese commerce. While the interaction is not the high politics of a deal involving Mrs.

Thatcher and Mr. Muldoon in New Zealand, there is no question that the bureaucrats and others who decide the policies of Japan have a strong influence on both JAL and ANA.

The King of Morocco

Another series of interesting meetings with political people and appointees that comes to mind occurred while I was on a world tour of the DC-10 in 1972. During that tour I met many high-ranking people. One stop was in Morocco. We had arranged a demonstration flight for the King of Morocco and his chief pilot. The pilot, it turned out, had been the pilot of the king's B727 when it was attacked by Morocco's own fighters in a coup attempt. The pilot saved the king by safely landing the airplane after it had been shot at and returning the king to the palace. In recognition, he was then made the chief pilot of the air force. As a result, the pilot had quite a bit of influence. After showing the airplane to Royal Air Morocco, we had the king join us on a flight with his chief pilot flying the airplane. I'm always amazed how pilots can fly these airplanes without much familiarization, but I'm also always pleased to see another test pilot from McDonnell Douglas or Boeing sitting beside them when they are allowed to do it. Anyway, the flight proceeded, and the king wanted to go around again. As I remember it, the king himself was now at the controls. GE had previously arranged a luncheon for the king and the U.S. Ambassador to Morocco to take place after the flight, but we were about half an hour late when we finally rolled up to the lunch location. As we walked into the room, everyone noticed that the ambassador, who was supposed to represent the United States, was sitting there eating his lunch—halfway through in fact. I went over to him and asked him why he had not had the courtesy to wait for the king and the rest of us. He said he was too busy to worry about the king. I could only shake my head and consider that, if his attitude was typical of our ambassadors, it was no wonder that we get a reputation for arrogance around the world. I also noted in my travels that most of the U.S. ambassadors I met—with the exception of one who had emigrated from Germany—really had very little business sense and seemed far more interested in enjoying themselves and having a good time in these countries than helping our economy. I did meet a good U.S. ambassador in Singapore. Unfortunately, GE never achieved very much the whole time we tried to sell in Singapore—and still have not. Pratt & Whitney, on the other hand, had a better ambassador by the name of Ambrose Young who facilitated their orders.

Bush the Elder

Just after the first Gulf War, I got a call from my office in Cincinnati that President George H.W. Bush was coming to town to make a speech and would like to visit the plant. This gave us all a big thrill because the United States and its allies had just won the Gulf War, and Bush seemed to be riding high, and we gladly accepted this honor. A lot of our Evendale facility is two-story buildings, but much of it is underground. The Secret Service came with their dogs and sniffed all of the cabinets of the employees. I suppose they were sniffing for drugs or explosives—

With President George H.W. Bush touring the Evendale plant.

I'm not sure what. We had to shut off the bottom floor of the building while President Bush and his entourage were inside. It turned out to be quite an interesting visit. My staff and I gave him a presentation about GE and what we were doing in Aircraft Engines. He listened very closely and asked some very good questions. Obviously he was very interested in airplanes and had been very well briefed. He was very congenial, and compared with all of the politicians I had met, he was certainly one of the brightest.

Generally, I find talking with politicians something like talking to pieces of wood, but not so the senior George Bush. Along with him, however, was Bill Gradison, our local congressional representative, who was running for reelection. I was absolutely amazed at the lack of interest Gradison showed in our employees. These were people who had voted for him, and he did not even offer to shake anyone's hand, whereas President Bush was freely doing that. People said, "Well, Brian, he didn't want to upstage the President." That's baloney! He was always 10 to 20 paces behind Bush and could have easily shaken some hands and endeared himself to our employees. He probably would have picked up a few votes as well. It turned out he was reelected anyway but decided to retire after the election—so his good will would not have mattered one way or another. I was really disappointed in both of our local representatives. Neither Gradison nor Tom Luken did much of a job for us at GE. We had many political issues to deal with on various engine programs and, as much as I hate to say this, we received far more help from Ted Kennedy and Tip O'Neill and the people from Massachusetts where our Lynn plant is located. They knew much more about what the Lynn plant meant to the Massachusetts area than the two local congressmen knew about the impact of our operations on Cincinnati. Of course, the worst representation we had was from Ohio's Senator Howard Metzenbaum. Metzenbaum was a strong proponent of the State of Israel in the Senate. We were making engines for a number of Israeli aircraft, but—to our disappointment—Metzenbaum failed to support GE's operation in his home state. Our anticipation was that to some extent "my friend's friend is my friend" would operate. Instead, a sort of opposition to big business prevailed. He just didn't want anything to do with us. I never quite understood this.

In any event, the presidential visit went very well. At the end of it, one of our people who had designed a very nice stamp depicting 50 years of jet engine development (and the running of the first jet engine) formally presented it to President Bush. We would have to apply to the appropriate people at the U.S. Postal Service, but we had George Bush sign it endorsing the stamp. In the end, our stamp lost out to Marilyn Monroe. That really upset me. That showed me again that the bureaucrats sometimes control more than the politicians do.

The U.K. and British Airways

An interesting highlight of our sale of the GE90 to British Airways was that we were really concerned that, no matter what the airline wanted, the decision would be made by the British government since they had put a tremendous amount of money into supporting Rolls-Royce's Trent engine. Their government was very proud of what Rolls-Royce had accomplished, and although we have had our ups and downs with Rolls-Royce, we never really put any political pressure on anyone in the United States to have them stop selling engines here. We just did not think it would work. We also did not think it would be in the best interests of the free-enterprise system. It was pretty clear toward the end of the competition, however, when BA was really close to selecting us, that they had to make some presentations to the government. Of course, at that time we were also negotiating to buy BA's Wales engine overhaul facility. Our purchase would be a big benefit to the Welsh people and the Welsh Development Authority because they knew that BA wanted to be out of the overhaul business and that we would expand the facility more than BA ever considered. We had a very good representative consultant, Sir Gordon Reece, who worked very closely with James Barrett, our man in England. They did the rounds to all of the political wheelers and dealers and informed them we were planning to build up our engine service business in the United Kingdom, as well as providing the best engine to BA. Even though we had done a good job on that front, we were concerned that as the decision got close, Prime Minister John Major would stop our deal. It turned out that we must have done what was right, but most importantly, British Airways' Lord John King and Colin Marshall had done their homework. As I described a bit earlier, they

Politics and the Engineer 161

eventually made the decision in our favor. Of course, we ended up buying the facility in Wales as well. At the time the facility was handed over to us, which was just before the general election in England, John Major came down to Wales with his wife for the ceremony. I found them both very nice and very pleasant people, although I felt that he did not have quite the zip I expected from a politician. He did seem to have his feet on the ground, though. He won that election—the first election after Margaret Thatcher had resigned—but he was eventually unseated in 1997. Later, when Tony Blair became prime minister, Lord King and Sir Colin Marshall essentially left control of British Airways in the hands of Bob Ayling, who had come from the financial department of the British government. Ayling had a pretty close connection with Blair, and the Rolls-Royce guys decided they would work this relationship so that GE would not sell in their backyard again. British Airways' next buy of 777s did not have our engines. This was obviously very disappointing for us. We really believe our engine served them in good stead, has not let them down, and continues to do a great job for them. I can only hope they will buy the long-range 777 with our engines when it comes up for purchase.

United Technologies

After I retired and just after I had my heart operation, I was asked by GE to represent them on a business tour to Southeast Asia being headed up by George David of UTC, and I flew there with several other businessmen. Originally we were supposed to fly on the Sultan of Brunei's airplane. We eventually ended up on an old Boeing 707. I must say the airplane was not the most comfortable thing in the world, and having just recovered from major surgery, I felt a bit stupid agreeing to be on the trip. However, it was a very interesting trip because I met a lot of politicians in Southeast Asia. I was very proud to represent GE and to note how highly regarded GE people were in the various countries we visited.

A most remarkable meeting was held for the prime minister of Malaysia. Our group was seated together in a room, and we were asked in turn to tell the prime minister who we were and what we did. General Alexander Haig was there. When his turn came, he stood and said, "I'm the only man who has been President of the United States twice, unofficially." That drew a big laugh from everyone. I found it all very ironic. There is no question that sometimes UTC does an excellent job interacting in the high levels of politics. There is also no question in my mind that Haig and UTC's Harry Gray exerted a major effort to stop GE from having what they felt was too great a market share in military programs. I am convinced that if both Pratt & Whitney and GE engines had been allowed to compete on pure merit, we would have won a lot more of the orders!

Environmentalists

When jet engines first appeared in use, people heard the noise and saw the smoke as evidence of power. After jets entered commercial service, the fascination with—or even acceptance of—noise and smoke waned. High-bypass engines, with less high-velocity gas coming out the back, reduced noise dramatically. Smoke, of course, is the result of inefficient combustion, and as engine designers we wanted to solve that problem. By the time we created the CF6 for the DC-10, we felt confident enough to stipulate in the contract that no smoke trail would be visible from this engine—and none was. I think we always did better at this than our competitors. The military was not particularly concerned about noise and smoke, but as combustion technology improved, the smoke from military engines began to disappear as well.

We were pushed to do much of this environmental improvement through federal or international aviation regulations, either actual or anticipated. When we were designing the engine for the Douglas DC-10, we were told that, unless something was done to improve the combustion systems on Boeing's 727s, all those airplanes would be out of the sky before the DC-10 was five years old. The DC-10 went into service more than 30 years ago, and 727s are still flying with trails of smoke coming out of their engines. The airlines could not afford to ground them, and their pressure on the regulating bodies continually postponed any rulemaking that would have required them to do so.

Meanwhile, we kept delivering on promises of ever-improving combustion. As our standards went up, environmentalists believed we could do better and better. Absolutely perfect combustion of perfectly pure hydrocarbon fuel in the best of all possible worlds would yield carbon dioxide and water. We were good, but by no means perfect. We could not get to combustion that produced just carbon dioxide and water as by-products, but the environmentalists were already arguing that we should eliminate the carbon dioxide from the equation, I guess by towing trees behind the airplane.

In truth, as pressure ratios and temperatures got higher, we got visibly cleaner combustion but produced more nitrides of oxygen, or NO_x, which are greenhouse gases. We are working to reduce those by adding a second ring of nozzles in the combustor and introducing specially designed swirling patterns into the combusting gases to burn everything we can before it goes out the back of the engine. Technology is getting close, but *perfect* will probably always elude us. Some nontechnical people seem to have trouble understanding that.

The other issue environmentalists and others raise is noise. (It continues to surprise me that people will buy a house in the takeoff or landing pattern of a major airport—usually at a lower price because the property is near an airport—and then complain that the aircraft noise is ruining the quality of their lives.) Aircraft engines have gotten quieter and quieter, and the new one's are almost undetectable on landing. Unfortunately, airplanes rushing through the air, especially airplanes with flaps extended and gear down for landing, are not quiet. People listen to the sound, and they think they are hearing engines. I am afraid there is little we can do there, and the airframers may not have much latitude either.

On takeoff the engines are doing the most work they will do during the flight and are making the most noise they will make. Big, relatively slow moving fans and acoustic treatments in the nacelle help minimize that. (There still is something like a buzzing sound that comes from the fan and can be heard in the airplane on takeoff. The engineers are working on that one.) We have also introduced exhaust nozzles that look something like the petals of a daisy. These nozzles mix the exhaust gases to reduce the noise they produce. Frankly, we have gone about as far as we can on noise reduction.

Airport noise is measured at various distances on the ground from the runways. The resulting pattern is called the noise footprint. Each airplane-engine combination has a different footprint. One way to reduce the size of that footprint is to get the airplane up and away quickly. The Boeing 777 has two powerful engines. The engines are powerful because they are designed so that one engine could power the plane if the other should fail. With this extra power, the 777 can get up and away quickly, and the noise experienced on the ground is close to the airport, perhaps even inside the airport boundaries. A fully laden 747, by comparison, will take a longer time climbing under full power, and so its noise footprint may extend well beyond the runway.

In short, we have resolved most of the environmental issues involving aircraft engines to the extent that physics and funds allow. Unfortunately, economics demands that older airplanes and older engines keep flying. Frankly, we would love to sell new engines to replace them, but we would much rather see our airline customers return to profitability first. That may mean that the air may not be as clean or as quiet as some activists would like for a while.

CHAPTER 15
My Sporting Life— Some Side Thoughts

I became interested in sports during World War II when I was about nine years old. At that time, there was, of course, no television and not much to listen to on the radio. For amusement, all of the kids in the neighborhood played phantom war games that consisted of building small forts out of the bricks left by builders who had stopped working on new houses because of the war. As I mentioned, my involvement in this game came to an abrupt end when one of the enemies in an opposing fort threw a small rock into our fort, accidentally hitting my sister on the head. For forts and rocks, we then substituted rival soccer games.

Soccer

In those days in England every neighborhood had a soccer team. There were no leagues, and so we had to arrange our own matches and provide a ball, which was hard to get in those days. For some reason, this task fell on my shoulders. We had some rough matches, and without a referee most of the time, they often got pretty heated. I was a big kid, and at first I played goalkeeper, but I found that was not much fun, as I didn't get much exercise. As I found, there was another disadvantage to the goalie's role. If your team was inferior to the opponent's, as the last line of defense, you really took a lot of flack from your own teammates.

After I started playing fullback and halfback, I really started to enjoy the game. Looking back I am amazed at how many matches I played. Some weekends I played for the school, the Navy Cadets (I wasn't a member), and the Air Cadets. While I got bruised up, I cannot remember feeling all that exhausted when I went to school on Monday. I also started to follow professional soccer, and my team was Chelsea (also my dad's favorite team). My best friend's team was Arsenal, and I can remember some grand arguments as to which was the better team. Although Chelsea picked up some star players after the war ended, I will never forget a memorable match between Chelsea and Moscow Dynamo at Stanford Bridge. At that time, the stadium held 60,000 people, most of whom were standing. Somehow there were at least 100,000 packed in there for that match. At times when the crowd surged, we thought we would be crushed. Several years after that, due to similar overcrowding, there were several people killed in another stadium, and seats were added to most of the big stadiums, which made it safer and more comfortable but a lot less affordable. I think we used to pay 18 pence (about 25 cents) for a ticket. It is nearly $30 for a ticket now, if you are lucky. Times have changed and so has soccer, which is now played at a much faster pace and with a lot more skill, making it much more exciting.

Just after the war there was one kid who always tried to play with us. At the time we were all 14 and 15 years old. He was only 13 and much smaller. Suddenly, he had a growth spurt, not only in size but also in skill. He was picked as a schoolboy international when he was 15 and played for Manchester United and England. By the time he was 17 or 18, you can be sure he did not want to play with us any more!

I did play a while for the Watford Juniors. (Watford later became a premier-level team and was partly owned by Elton John.) We did well as a team, but all things considered, I was really too big to be a great soccer player.

Cricket

I also played cricket in the summer—again with pickup sides and with any equipment we could find or make. As with soccer, we never had any coaches, and we were pretty much self-taught by watching the older guys play. When I was 18, I joined the local Kenton Cricket Club and played for them for several years. We would start a match about 11:00 in the morning, have lunch at 1:00, tea break at 4:00, and dinner and a drink after we finished at about 6:30. Quite a civilized sport! Cricket is essentially unfathomable to most Americans, but it is loved in a lot of places in the world. The great thing about the sport is that you get to meet a lot of people in a relatively relaxed atmosphere. Lately it has become more vigorous, however, with bowlers who bowl so fast that batters have to wear face protectors and padding on the rest of their bodies in addition to that on their legs.

Basketball

As a youth, when I went away to college in Newcastle, the weather was not very good for tennis, even though during the fall and winter sessions I did play for the university a couple of times. I really wanted to play soccer for the university—and stood a pretty good chance until I broke my ankle during my first game of basketball for the university. (I was only selected for the basketball team, I am sure, because I was 6'4", and most of the team was smaller. I had never played basketball before.) After my ankle recovered, I played a lot of basketball for the college. We were unbeaten until we went to play Edinburg University. We went up from Newcastle by train and arrived just in time for the game that was to be played in a gym about twice the size of the one we played in. We were warming up when the Edinburg team entered. They were all about 6'8" or taller—they looked like giants—and most of them were from the United States on an exchange program. This first sight proved ominous, and they beat us soundly. I don't think we got more than 20 points, and they must have scored 100. Luckily for us, as with all university sports in the United Kingdom, there were very few spectators.

Swimming

At the time I was playing basketball, I also played water polo and swam on the swim team. Water polo is a pretty rough sport, but it also let us travel around the country to play. In one game against Leeds University, a fog had settled over the pool and was so thick that we could not see from one end to the other. It amazes me that this memory is still so vivid today that I could close my eyes and be there.

I learned to swim when I was about 6 and loved it. I was fairly good at it and was asked when I was 13 or 14 if I was interested in training in a nearby town to try out for the Olympics. After several sessions of training—plowing up and down the pool for what seemed like hours on end that no doubt improved my times—I decided I would rather play a team sport where at least I could see the opponents. Being part of a team was more suited to my personality.

Tennis

My tennis career started when I was about 20. One of my friends, Bill Moxon, had taken it up and needed someone to play against. I found that because of my height and mobility, even though I was heavy, I could play with the best of them. I also found this to be the most enjoyable sport, especially when I played against people who were equal to or better than me. I hated playing against slow ball players.

Without tennis I almost certainly would not have met Jill. I was playing in a mixed-doubles match with another girl as my partner when I smashed the ball and hit my mother-in-law-to-be soundly. Jill came over to see what all of the fuss was about, and I asked her out the next day. I guess it was a lucky smash.

I carried this sport into my adult life, and tennis helped me make some great friends outside of GE in Cincinnati. With all of my traveling, it was difficult to have a consistent social life, although my wife, who was a great tennis player, made a lot of good friends. Her tennis connections also helped me to meet people outside of GE.

When I returned to Cincinnati in 1967 after my stay in Lynn, I had two conditions for the real estate salesman who sold us a new home in Cincinnati. The first was to get into the local swim club, and the second was to get into the private indoor tennis club. He came through on both of these, which was great for my family and me, as my kids were very active in the swim teams. I made friends with some men who were about my skill level at tennis, and we decided in 1968 to form a Sunday morning tennis group. We had a scheme for getting on two courts at 8:30 every Sunday morning from October through March, and each person would play three matches with one of the others until we rotated through the whole group. This took some scheduling and a good set of records, which eventually became much easier when computers became available. Initially we called it the Masters Tournament, but in the end, when we all started to get long in the tooth, it became the Grand Masters. Each player was responsible for getting a substitute who would show up on time on Sunday mornings if he were unavailable. Because I traveled quite a bit, I used to have a better record, because my substitutes were usually young Super Stars. This really upset some of my fellow Masters. We then put in a handicap system that was based on the previous year's results, although you weren't allowed to use your handicaps when you had a substitute. This helped make things a little more even, but it was not a rule I favored too much.

At the end of the season, we had playoffs in which the top points maker got to play with the lowest (No. 8) and No. 2 would play with No. 7, etc. We all put $20 in the pot at the beginning of the season, and the top points getters shared the spoils. I don't believe I was ever a big winner, but I may have broken even. We did have a lot of fun, however, and it gave us all a regular game. The most important part of Sunday morning after tennis was the half-hour we spent together discussing the world's problems and, more importantly, telling the latest stories we had heard. (Most of the stories were not the kind you would repeat in refined company.)

Golf

I always felt very lucky that I could pick up any sport fairly quickly and become an average performer with my peers in almost no time. The one sport in which I would really like to go beyond average, but find myself stymied whether by lack of mechanics or concentration, is golf. I started to play golf with some seriousness while attending GE meetings. I would play golf one day and tennis the other. (I did well at tennis but not too well at golf.) I started with a 26 handicap, and after many years of playing, whittled that down to 15 to 17 depending on the course. Unfortunately, I have never improved from there. I really admire people who can consistently shoot below 80. I'm afraid I will have to play many more years before I can shoot my age!

It is golf, however, that has given me the most pleasure over the last 20 years. I have played many great courses with many fine people. I have enjoyed most of the major courses around the world including Augusta National, Pebble Beach, Pine Valley, and St. Andrews, as well as courses in Australia, New Zealand, Thailand, Indonesia, Japan, and France. The most enjoyable rounds involved trips to Scotland when we took several of our friends from airlines around the world, who all seemed to really enjoy playing golf there. Since retiring, I still organize a group of old friends including Hollis Harris, Ed Bavaria, Dick Smith, Bob Turnbull, Neil Armstrong, Bob Hawkins, and our sons on a biannual trip to Scotland. While working for GE, I was lucky to be able to dovetail these trips with a visit to London to watch the tennis finals at Wimbledon, thanks to Bob Wright at NBC who provided the tickets. I have been lucky to have seen some great finals, including the classic match between Borg and McEnroe.

Lessons

As a by-product of playing all of these sports, I have also become an avid sports spectator. I find watching a sports competition really relaxing, as the outcome is a true result, generally without the influence of political prejudice or control. Throughout my career at GE, I spent most of what little time I had when not working—which took up most of my time—either playing or watching sports. I have learned to enjoy American football and sometimes wish I had spent my youth in the United States so that I could have played it. It looks like a lot of fun. However, I am sure that, if I had played, I would have even more aches and pains in retirement than I have from the sports I did play.

When I was running the Engineering Department in 1976 and did not have to travel as much as in my previous jobs, I took time several evenings and weekends to coach high school soccer, which was just becoming popular in the United States. We had a pool of boys who had no idea about the game from which to pick our

My Sporting Life—Some Side Thoughts

teams. I decided to have them all race around the field and picked those who could run the best. I felt I could teach them the elements of soccer but could not teach them to run. As a result, our team won its league without a loss. Unfortunately, I had to take a trip the day of the cup final, and our boys lost 2-1, a big disappointment for us all. I would like to say we lost because I wasn't there, but that was not the case. The opposing team was just a lot better, even though our team played really well. As a result of my involvement at the youth level, my son, David, also became interested in soccer and played on the team.

I have learned a lot from my participation in sports. Team sports help you to understand that, no matter how good you or any other person on the team might be, individuals cannot do it alone. That is true not only in sports but in business as well. In both team sports and business, you need to be able to pick good people who can work together. This happened for me at GE.

The most important lesson from sports, however, is the one they teach you about losing. Losing lets you learn where your weaknesses are, so that you can improve yourself and your team. There is as much to learn from your losses as from your wins if you do not dwell on the losses but on the lessons. All this being said, however, winning is definitely better than losing.

Chapter 16
The Conquistadores

In 1976, I was invited by Ed Hood to attend a Conquistadores meeting. The Conquistadores del Cielo is a club originally formed by a group of aerospace pioneers. It began in 1937 when Mr. Frey got his first transcontinental flight certificate at TWA, and in the beginning it consisted mainly of airline executives. Frey thought it would be a good idea to bring his friends together for a regular social gathering. The Conquistadores would have a three-day meeting at a remote ranch once a year where they would ride, shoot, and hunt. Eventually the group got into tennis, golf, and other more civilized sports, but originally it was all ranch-type activities.

Although they were held at several other places in the early years, most of the fall meetings, which are the important ones, take place at the A-Bar-A Ranch in Encampment, Wyoming. The drive to the A-Bar-A from the airport takes you through the beautiful country of Wyoming with its big blue skies, and since it's at 7000 ft, it seems as if you can see forever. After you enter the ranch gates from the highway, you still have a seven-mile drive through the mountains with streams on one side of a rugged road. Finally, you arrive at the main entrance surrounded by green fields, a scene that is just like sitting in the bowl where you could imagine Shangri-la was. It's a fabulous area of the country with clear, clean air and pure water.

I considered it a great privilege to be invited to this meeting, especially as I was attending a fall meeting, a requirement before being invited to become a member. I was very lucky to be elected to membership in 1977. This was a very memorable occasion for me as membership in the Conquistadores reflects on the standing you have in the industry. In my early days as a member, many of the old airline pioneers were still attending the meetings, and it was really fascinating to hear their stories and learn how they got started in the industry. Of course, there were also a lot of pioneers on the industrial side, but people from outside the airlines were not invited to join until three or four years after the club was started.

Risking life and limb

The Conquistadores is a meeting of friends and acquaintances, and we all go out of our way to avoid business discussions, primarily for social reasons but also for the fact that talking business in this setting could possibly be construed as violating antitrust laws. These get-togethers really are opportunities to meet your business associates as well as your competition and see how well they do in a variety of sports. You can get a good feeling for people when you see them compete in tennis, golf or shooting. What impresses me is how skillful most of them are in a wide variety of sports. I'm fascinated and amazed to see some of these guys participate in a rodeo every year and watch how willing they are to risk life and limb charging around on these big horses. It's quite a spectacle. Interestingly, some of them had never ridden a horse until they joined the club. I can honestly

Five horsemen at the A-A Ranch: Rheinhardt Abraham, Lufthansa; me; Adam Thompson, British Caladonian; Harry Stonecipher, GE/Boeing; Jean Pierson, Airbus.

say I had never ridden a horse before I became a member, and I made sure I always had a very strong, slow horse when I started. One day I was riding with a group that included Giovanni Bisignani, the CEO of Alitalia at the time, who was a great horseman. We were riding along the trail, my horse trotting along behind his, and all was going very well. I was feeling pretty comfortable—as comfortable as I could feel on a horse—when all of a sudden he or his horse decided that instead of walking down the side of the stream and up the other side, he would just jump over it. My horse decided the same thing. I thought the end of my life had come as I was flying through the air. I stayed attached to the horse, but just barely. I couldn't walk very well for a couple of hours after that.

Selective membership

The number of beds available at the ranch determines the size of the club, and so the members have to be pretty selective as to who is invited to join. Of course, you also have to be selective about whom you share a cabin with since some of the members, particularly the older ones—probably including myself—snore loudly. If you get the wrong cabin mate, a good night's sleep is almost out of the question. One old chap, Al Adams, was a really great Conquistador. He had the misfortune to share a cabin with another good friend, Charlie Forsyth. Al was probably 20 years older than Charlie, but Charlie had the most amazing trumpeting snore you could ever imagine. I'll never forget Al coming to breakfast complaining about sleeping in the same room with the man with a horrible snore. He ended up sleeping in the bathtub with his ears covered in an attempt to avoid the noise.

The Conquistadores provided a good opportunity to meet some potential customers as well as assess your competitors. It's very interesting how friendly you can become with your competitors—and sometimes even see good in them. (I hope they think the same about me.) My main sport when I first joined the club was tennis. I was very lucky to win the tennis competition with several different partners in 1977, 1981, and 1983. One tennis partner was John Richardson, who worked for Hughes. He was a good player, and we played well together. Not only did we win once, but we were also runners-up twice. The rules of the club are that the winners of any tournament—whether it be horseshoes, pétanque, or whatever—has to run the tournament the following year. So, if you win regularly, you can only compete every other year. John and I won, then missed a year, then were runner-up two years in a row. John, unfortunately, passed away, and I had to find a new partner. K. Hurtt and I then won in 1981 and 1983. Once you've won a tournament three times, you are not allowed to win anymore, and so I retired from official competition after 1983. I still played quite a lot of tennis, even though playing tennis at 7000 ft is quite a different experience from playing near sea level.

Another sport I won at was pétanque. Pétanque is one of Europe's most popular outdoor games. It is a distant cousin of horseshoes and a close relative of bocce. The aim is to toss or roll a number of steel balls as close as possible to a small wooden aim ball. Players take turns, and whoever ends up closest to the aim ball when all of the balls are played wins. Unlike horseshoes, where the aim stake is fixed, pétanque's aim ball may be hit at any time, which can completely turn around the score at the last second. Whereas the official bocce rules call for a prepared court with markers and sideboards, pétanque can be played on most outdoor surfaces. In 1990, I teamed with Jean Pierson, head of Airbus Industrie, and we took the championship. Jean was an expert in pétanque in France where they play on gravel. We play on grass, which is not quite the same game, but his skill brought us through and we won.

Pride in leadership

One of my greatest experiences in the club was to help lead it, first as a member of the board of directors and eventually in the position of first president and then chairman. This was a great privilege, and the outstanding support I got from GE people really helped me do the best possible job. Our spring meeting during my

Captains of Industry: Sunning on a beach in Tunisia with Jean Pierson.

presidency was at The Wigwam in Arizona. Part of the responsibility of the president is to organize the spring meeting and have an annual photo book printed. I had tremendous support from Wayman Brown in Aircraft Engines' marketing department. Our objective was to make the annual book equal to any done before at a lower cost, and we managed to do that through Wayman's artistic skills.

The camaraderie of the club is something special. We have a lot of good people. Not only are there those highly qualified individuals who are or were chairmen or presidents of their companies, but we also have fun-lovers who are simply great story tellers and magnanimous people we all enjoy being around and listening to. There is also a group that really enjoys gambling, and so there are often some pretty high-stakes card games going on. I've never had the urge to lose my money that way. I consider that it took a long time for me to earn it and I certainly don't want to give it away through inept card playing.

The general attitude of the members is very cooperative, and if you have a problem, you can usually call on a fellow Conquistadores to help you out of it. That is very reassuring. We have also been blessed by having some very fine doctors on site at the ranch. As you can imagine with people riding horses, playing tennis, and other strenuous sports, we have one or two injuries every year. One such person was riding down the trail and his horse slipped. He fell into a ravine and hit his head. Luckily for him someone was very close and pulled him out, otherwise he would probably have drowned. He did suffer quite extensive injuries, and full recovery took several weeks. As you can imagine, it's extremely embarrassing to return to work all battered and bruised from what is supposed to be a couple of days of vacation.

A great honor was bestowed on me in 1997 when I was awarded the Big Horse, a trophy given for being a good all-around wrangler and a good sport who contributed to the club. I felt very honored to have been selected for this special recognition. In fact, I've felt very honored to be a member of this club for 25 years. While I was on the board and chairman, my objective was to lower the average age of the club members. As I've said, the club had to be very selective in who is chosen for membership. My main worry was that the average age was going up, and if we weren't careful, we would all be too old to enjoy any activities together. I did convince the Conquistadores to increase the number of aspirantes, or new members inducted each year, and generally that has revitalized the club to some extent.

Sometimes, the people who have struggled to make it all of the way to the top in their businesses seem to be pretty worried about their status. That can make them stiff and joyless. I wanted to see the Conquistadores expand by adding more younger people—good story tellers, just plain guys who might not be president but who are or will be at a high level in their companies—so the rest of us can continue to have fun. This is now happening, and we are getting some great new Conquistadores.

CHAPTER 17
Values in the Turmoil of Leadership

I have been successful in a technologically complex and operationally convoluted business. I have gained respect and admiration in a demanding industry. I have made a lot of money. Some might see this as an end worth achieving, but they are wondering, why you, Brian? What did you do to make this all happen? From the perspective of that boy looking into the 1940 sky at a faltering Hawker Hurricane heading in his direction, all that has happened is well beyond my expectations—and certainly beyond my conscious planning. When I look back, however, there are some things I did do over the course of my life that created the options and opportunities that, when acted on, led to where I stand now.

Obviously, upbringing has something to do with where you choose to go and where you do not. Although my mother and father were not religious, they were really ethical people. I do not think I ever saw my mom or dad do anything nasty or vindictive, and I am sure that some of that rubbed off. Then there are the more dramatic, life-changing events that caused me to be more interested in some things—such as airplanes and sports—and less interested in others. Finally, there are three things I might label: industriousness, organizing things and ideas, and working with people. Another way of describing these is personal skill and energy, management ability, and the desire to lead. How I came by these I do not know, but I am sure organizing sports and leading a Boy Scout patrol helped.

I have always been fascinated by how things work. As a boy, seeing an airplane lift off the ground was utterly intriguing. That wondering curiosity never faded. In fact, it grew as I understood more about the process of making things fly. I became an engineer. I think that, even today in this e-commerce world, with an engineering background, you can do anything you want to, because engineering teaches you about processes—how things work in a step-by-step fashion. Once you know that, you can become a finance person, or a marketing person or a manager, but finance, marketing, or management will not pave the way to becoming an engineer.

On the other hand, engineers tend to get locked into the things they really enjoy doing. They see beauty in esoteric subjects like aerodynamic analysis. They make computer tools to help them do a better job, and they find playing with those tools to be extreme fun, just like video games. They are taught to think in logical steps, and the parts of life that defy logic often elude them. If I am any example, engineers have trouble tolerating politics and other things that conflict with mathematical reasoning. Sometimes this can be limiting.

If anything in my life helped me overcome those kinds of limitations, it was liking people. Anyone who achieves anything in the world today must do it with the cooperation and skill of many other good people. Selecting the good ones is considerably easier if you actually find associations with your fellow humans a pleasure. During my career as a leader, many of my bosses and colleagues accused me of being too soft on the people under me. I found that it was not valuable to be hard on people until they failed to perform. This is not something I learned from some course on management. In fact, while I often gave lectures on managing people at GE's Leadership Course, I never actually attended such a course myself. Somehow, I seem to have inherited some instincts that made people want to work with me. Beyond that, I have no explanation.

Another difficult-to-define attribute that I think is essential for any real success in business is the ability to withstand stress. This may be something one inherits as well, although I am not completely sure about that. The flipside of this is a quality I would call bravery. Bravery demands a stockpile of conviction that makes a person believe that he or she can take the next leap in technology—or the next step in his or her life—not collapsing in a feedback-loop of worrying, but getting on with it and doing it instead. After bravery starts working, it fills its own stockpile of conviction. How you get the initial deposit, I do not know, but I think family life has a lot to do with it.

I also watched the world around me and learned from it. I found that a number of people who worked for me had pre-plotted their lives. They even had their position and salary expectations defined for the future. In general, these were very frustrated people. Few ever accomplished their long-term goals. It seems that they failed by not being aggressive and open enough to take the big steps and make the big moves that are necessary to be a genuine success in business. Everything in my experience told me that, to achieve what you really want, you must expand your perspective to include all of the possibilities. While it is crucial to have goals, it is even more important to be prepared to go beyond them at a moment's notice. Continuously working hard with the resources you have is also important, as is concentrating on your present job to be as successful as you can.

How to be lucky

Luck seems to be a valuable part of the equation of success as well—although some would deny its existence. In fact, it may be more important to believe you are lucky than to have any real evidence of it. A confident person is a "lucky" person. A person who believes in his luck will be energized by setbacks and will get back into the fray with renewed vigor. A person who says, "There is no such thing as luck," will be careful and slow-moving in the face of adversity.

A lucky person loves to show off his luck by competing, and a person who acts as if he cannot lose usually wins. (I am sure that is why there was such an agenda of competition among members of the Conquistadores.) I had no ambition to business leadership as a low-level manager, but when I became head of Engineering at Lynn, I found myself saying, "These guys aren't so smart; I can do better than they can." After several promotions, I became a vice president and head of the Commercial Engines Operation. At that time, Gerhard Neumann was thinking about retirement. Jim Worsham and Bob Goldsmith were waiting in the wings to take over. Neumann said, "When I retire, one of you two will be the boss." I smiled to myself. I had had opportunities to run Rolls-Royce and other British companies and turned them down. I had never had a burning ambition to lead Aircraft Engines until it looked like somebody I did not respect was going to get the job. "To hell with watching from the sidelines," I thought. "If they could get it, why shouldn't I."

I then made sure that both Neumann and his superiors knew that I was keen to get the job. I also made an extra effort to be as cooperative as possible with my peers—even the ones who were my competitors—to make sure they knew I was a team player. When Fred MacFee was put in charge after Neumann left, Fred asked me to coordinate the efforts in Evendale while he oversaw Lynn where he was stationed. While not formally appointed to anything other than my job as vice president of Engineering, I was able to work with the rest of the team and demonstrate that I could handle many different types of problems. Most importantly, during this period I showed that I had a vision for the business, especially the commercial engine business where GE's greatest potential for the future rested.

Family support

I also went out of my way to expose the people who worked for me to the different facets of the business, including them and their wives on coordinating trips to our partners and customers. This helped show the spouses what GE Aircraft

Jill with some Chinese hosts during one of many marketing visits to China.

Engines was all about. Jill was very helpful in all of this. She had always been a great tennis player, and that helped immensely in the social end of business. More importantly, however, she had a sense of worldly wisdom that grew from early childhood trips with her parents. She never acted like the queen bee and was able to get on well both with the wives and their husbands. Jill and I were seen as team players that people wanted to work with.

Jill has been a blessing to me in many ways. First and foremost, she did an outstanding job of bringing up our children, who matured to be marvelous adults. This helped make my commitment to GE a lot easier. For my part, when I was on the road, I did call every day, no matter where I found myself. I think that added a feeling of stability in this fast, ever-changing world, but Jill was the one who made the family tick.

While my job brought her in contact with many people, Jill was always her own person. She was always herself whether she was talking with Prince Phillip, Ronald Reagan, Reg Jones, Jack Welch, or the wife of a factory worker at an open house. Jill was respected for herself, not because she was my wife.

Having a supportive wife and family was a great incentive for me to work hard to make our family life better. Without Jill's cooperation, our immigration to the United States would not have been possible. She made it very easy for me and adapted well to the United States. That trip to America for a short two-year working vacation proved to be our biggest piece of luck. Who can say how things would have worked out for us had we stayed in England?

All I can say is that the opportunities offered by GE and the acceptance that Jill and I felt coming to the United States generated a great life with many adventures and many advantages.

Philosophy of life

I note that people seldom live by philosophy. Instead they discover certain patterns after doing a good bit of living and, with some mental legerdemain, see a plan and intuit an intent. To say that my philosophy of life is "do the best you can with what you have and take every opportunity that presents itself" does not really say much. In truth, I am pretty much of a fatalist. I take the next step, and then the next step, and then the next step.

I loved the work I was doing, and that passion is a big part of my experience. I did not think about retirement or amassing the wherewithal to live well after GE until I was well over 50. I invested my incentive compensation in GE stock, and it did well. I trusted in a good pension program, and I was right.

I am sure that Jack Welch thought I was often too focused on the Aircraft Engine business and not focused enough on making money for GE. Perhaps he was right, but I was not stupid. I knew that we had to make lots of money to build the stable of commercial engines we now have in the marketplace. And we did. This is not something I could have done alone. I was given great people to work with, frequently people much smarter than me. And together we did this. Part of my job as a leader was to steadfastly represent all of those people in my negotiations with Jack and his finance people over operating revenue and the cost of our development projects. Did Jack and I disagree? Yes, we frequently did, but even in

Values in the Turmoil of Leadership

compromise, I still remained true to the team. I think the people who worked for me knew and understood that. As a result, they would loyally follow me as I charged up the hill of our next technological or financial problem. It was all very exciting.

This consuming passion for the business takes its toll, however, and I did not realize all I had given up until after I left the fray. It seems that every Saturday I was arriving home after a tiring trip, and every Sunday I was packing up to go again. In fact, I was traveling about half the time, visiting with customers, partners and field representatives and working on the way back and forth. We used to have customers come to our house, and people would say, "Oh, isn't that nice." Nice, yes, but you are working all of the time you are with customers. Jill raised our children, and when they were grown, I realized I had missed a large part of an important experience.

Chapter 18
Life After GE

As my retirement approached, I focused on applying much more effort toward getting closer to my family. As I mentioned, much of my 35 years in the business had been spent traveling the world in search of customers and partners. In addition, my position also demanded that Jill and I host many social visits from customers and partners in our home, which also took time and attention from the children. We did commit to regular time together as a family on Easter vacation, during the summer, and at Christmas, but in truth, Jill carried the load of helping the kids grow, and I sometimes felt as if I were a guest in their house—unless I was called on to be the enforcer in some family disturbance.

Today, the children are all grown and married. Linda is married to Henry Hernandez, and they have two sons, Kyle and Brian; Penny is married to John Dinsmore, with a son, Nick, and a daughter, Amanda; and David is married to Leslie Fleming, and they have two sons, Tyler and Colin. Our family that started with the two of us in 1957 is now up to 14. Nick and Kyle are at university, Amanda starts next year, and the rest are in grade school. Some are good scholars and some are good athletes—on the whole, a good American cross section. The Christmas before I retired in 1994, I took all 14 of us to St. Croix for vacation. Ten days before we left, Linda, our eldest daughter, gave birth to our youngest grandson, Brian, who is now nine. Time seems to have really flown. I guess that means we were having fun.

In my younger days, my father advised me always to buy more house than I thought I could really afford. I followed his advice and as a result, as I grew at GE, we ended up with very nice homes in Cincinnati, Key Largo, and Cape Cod. These are not only great places to live, but they also help the children and grand-

The Rowe family on vacation in St. Croix.

children visit quite frequently. These homes also support a number of sporting activities. Unfortunately, I found that golf, tennis, swimming, and boating, while very enjoyable, are not nearly as satisfying as working.

Apart from sports, I never really had a hobby in which I cared to invest much time. As a result, as I found myself facing retirement, I was concerned that I would not be able to keep as busy as I wanted to. Several people cautioned me not to become involved with any corporate boards as they felt I would soon become frustrated. I subsequently learned that they were correct about the frustration, but that this was offset by the satisfaction I felt from helping some good companies improve. In the process of doing that, I met some impressive people, and I also discovered that the world does not revolve exclusively around jet engines and airplanes. I learned that my experiences at General Electric helped me contribute in many ways. I am sure many people tired of me quoting chapter and verse about how things were done at GE—things like manpower development, financial control planning, and review cycles. To my surprise, however, they did adopt many of the ideas I suggested (although not as speedily as I might have hoped). All things considered, there is no doubt that GE was an outstanding teacher. I am pleased that I was able to pass on some of what I learned. Here are some thoughts about the boards on which I was invited to serve.

Fifth Third Bank

In 1980, shortly after I took over leadership of Aircraft Engines, Bill Rowe (no relation of mine, by the way) of the Cincinnati-based Fifth Third Bank asked me to become a member of that company's board. The Fifth Third Bank under Bill, Clem Buenger, and most recently George Schaefer has been tremendously successful. Fifth Third is recognized as one of the best middle-sized banks not only in North America but also by people throughout the world.

The most important thing I learned by being on a bank board was the necessity to keep an eye on overhead, as it can drain away the value of all of the good work that employees are doing. Fifth Third management, even though always young and vigorous, has been very successful at avoiding lending money to people who would not be able to pay it back and, at the same time, keeping the overhead down. Obviously, if you do not have to write off too much, your chances of making a profit improve greatly.

One of the big positive moves Fifth Third made was to issue stock options to their successful people. These stock options have done very well, creating a real incentive for the people involved. I found, however, that stock options could also have a negative influence from the team-building point of view, in that many of the recipients decide they can make enough money on their options to retire relatively early. As a result, it was hard to develop a satisfactory succession plan for upper management. George Schaefer was a very successful leader and relatively young, and so the people directly below him felt their chances of ever getting his job were slim. That accelerated their departures.

Since I left the board, pressure from the new members and some new government rules convinced George to formulate a succession plan. For my part, membership on this board was a great learning experience—especially the skill of patience.

Aerostructures Hamble Ltd.

As I was nearing the end of my career at GE, an opportunity came up to be a director of Aerostructures Hamble Ltd. in England. Lord King was the chairman. Aerostructures was a neat little company that built parts of several different airplanes and was a spinoff from British Aerospace of the United Kingdom. The company was led initially by one of the previous managers of BE Aerospace who had brought several of his relatives with him. Unfortunately, this nepotism got in the way of the company doing well. The company had quite a few financial problems, but these were resolved when we sold Aerostructures to another company. Lord King did a good job in holding the company together while going through the sale. Once the sale was complete, I resigned from the board and frankly vowed never again to take a directorship with a company overseas because of the toll taken by long-distance travel.

Stewart & Stevenson

While I was chairman of Aircraft Engines—about to step into retirement—Jim Stewart II, chairman of Stewart & Stevenson, asked me if I would like to join his board. He invited Jill and me to Houston to visit with him and the other directors. This is a very intriguing company that is mainly in the service business but also does an excellent job of making generating sets and pumps for both power generation and the oil industry. This company was really homegrown and had a lot of close relationships. It had only been public for about 10 years before I joined the board.

At my first board meeting, I found the company was in a dispute with the U.S. Air Force over some generating sets they were making for the Saudi Air Force. This issue hurt the company severely, even though it was over a very small amount of money and should have been settled quickly by a good lawyer. Instead, no steps were taken, and the matter dragged on. As a result, the CEO was forced to leave, and the company operated under an acting CEO for over a year, really suffering as a result. The lesson, of course, is that you cannot allow your business to get twisted over essentially small issues. You must act with a broad view of business realities in sight and fix the problems quickly. After a long period of incumbents, we managed to recruit a very good candidate for CFO, John Doster, a man with a GE background. Doster put in some effective systems and got them thinking about inventory, cash, terms, and all of the other things GE had been drilling into its businesses for several years.

The board eventually saw the logic in hiring a new CEO from the outside, and we were lucky to get Mike Grimes, also a former GE man, who has done an outstanding job. The truck side of the business was being run by a long-term Stewart & Stevenson person who was obviously not cutting it in that job. Doster found that Richard Wiater, a man who had worked for him in Aircraft Engines, was available. Wiater came in and did an outstanding job sorting out the truck-building business.

Atlas Air

In early 1995, I was asked to join the board of Atlas Air, a new cargo company run by Michael Chowdry, whom I had met through the Conquistadores. He was Pakistani by birth and had immigrated to the United States at age 25. His life really turned out to be one of those "it can only happen in America" stories. Michael graduated from the University of Minnesota, learning to fly while he was there. After graduation he became intrigued with flying and had many jobs in aviation. He then thought he would get into the airplane leasing business, which he did—successfully and unsuccessfully—for several years. Around 1990, his company owned one 747 freighter that was on lease to Pan American, which went bankrupt. Michael was driving through New York and, in response to this setback, said to the chap with him in the car, "I'm going to form an airline myself." As they drove past the statue of Atlas, Michael said, "We'll call it Atlas . . . and it will be a freight airline; and we'll concentrate on one type of airplane with one type of engine so we don't have a lot of relationships with many companies." This model, created by him, was called ACMI—Atlas provided the aircraft, crew, maintenance, and insurance—and the airlines, with which he worked out long-term leases, would provide the freight, fuel, and scheduling of the airplanes. Atlas quickly grew from 1 to 5 to 10 to 12 airplanes, and the business became very successful. It was one of the very few airlines to make any money in the last decade of the 20th century, and Chowdry did this while growing the company and developing the ACMI concept into something that was really booming up until the end of 2000.

Of course as the economy got worse, the freight business was one of the first to suffer. This is a different cyclic phenomenon than that experienced in the passenger airline business. Usually the airlines get hit about 6–12 months after there is a real recession. Airframe and engine manufacturers get hit about a year to a year-and-a-half after a recession. In the freight business the hit is almost instantaneous, because most of the air freight being carried is for just-in-time needs, and that segment shrinks and dries up immediately. Atlas was also shipping a lot of electronic components, and that sector dried up on its own at about the same time as the general recession began. All of this economic gloom happened just after Chowdry had built Atlas' fleet to 37 Boeing 747s—all with GE engines, I'm proud to say.

One of the biggest gambles I thought he made was when he decided he would buy new 747-400s. This was not a technical gamble, because these are great airplanes; it was a marketing gamble since Atlas had to find the people to use these airplanes. Michael traveled the world doing just that. He was a great salesman and a great negotiator with the airplane companies or the people with whom he was doing business. Unfortunately, an entrepreneur like Michael also ran a pretty loose-knit organization. It was always hard for the board of directors to know from whom in the organization, other than Michael, they were getting the straight scoop on business matters. Many of the people were working directly with Michael, sometimes on similar projects, and he was using them as checks and balances. It was always difficult for the directors to understand who was the check and who was the balance. Anyway, Michael was very successful, and the business

Life After GE

grew. While he divested himself of some of his shares when he took the company public, he owned approximately 50% of the business.

In early 2001 Michael decided that, rather than go to a GE technology conference in Hawaii, he would stay home in Denver to be interviewed by a gentleman from the *Wall Street Journal*. Unfortunately for both of them, Michael took the reporter for a ride in his L39 Czechoslovakian fighter-trainer, which he really loved. On takeoff, for some reason the plane crashed, and both of them were killed. This was a tragedy, not only for their families but also for Atlas as a company. I had been quite close to Michael, particularly on the technical side of airplane acquisition, and worked with him on trying to improve the management. He and I had become very good friends, as had our wives. As a result of the accident, Linda Chowdry, Michael's wife, who inherited a large percentage of the company, asked me if I would become chairman during an interim period to get everything organized and to help her select the new organization.

For me, this turned out to be a very frustrating and interesting challenge as I am pretty much a hands-on guy. To be chairman of the board but not the CEO proved quite difficult for me—not because I did not trust the people running Atlas but because I felt they had never learned to communicate properly with each other or the chairman of the board. As a result, it was hard to find out what was going on, although I did see improvement every week I worked with them. As the communication improved, we did a much better job of distributing the load of running the company.

I inherited this job in January 2001. As we approached summer, the freight market dropped by about 20% in three months. To keep Atlas going, we really had to dig in and implement some rapid cost reductions. We furloughed quite a few pilots. Not only did these layoffs conflict with the team building we were trying to do, but they also put more pressure on management. On the positive side, we got some management systems in place by consolidating all of the staff at one headquarters. We also began diversifying our fleet and bought Polar from GE so that we could help our customers at various load factors on long-range routes. The Boeing 747 is a long-range, high-payload airplane, and I think there is a great future for this application at Atlas, but there are also other demands in the marketplace.

We did not make the necessary cuts early enough. This hurt the total business. Running any business while forecasting its future is a very difficult thing to do. It is even more difficult to reassess the situation when things start to look bad if you are dealing with a company that had been growing at the rate Atlas was. They were encouraged to feel that the handwriting on the wall was not meant for them, saying, "Don't worry, it's going to get better. It probably won't last long." As a result, they were inclined to keep people on the payroll and airplanes flying unproductively too long. My objective was to convince them that they had to cut quickly. This turned out to be more difficult than it should have, but we eventually did it.

The relationship between the CEO and me was a little tenuous to say the least, but he had helped Michael create the company and as a result was the glue holding it together—and the rest of the board felt he was capable of running the company. We recruited a new CFO who we all felt would help us put a better financial organization in place—which he did. Late in 2002, while we were still suffering from the business downturn caused by the 9/11 tragedy and the Enron scandals, we decided to change from Arthur Andersen Audit Company to Ernst & Young. Almost concurrent with this, the new CFO and his team found that we had both inventory and impairment problems that should have been handled differently in earlier years. This caused further unrest in the business as it required that we restate earnings for the previous years. Of course we were also required to tell the Securities and Exchange Commission (SEC). This had a very negative effect on our stock but, at that time, did not affect the value of the company. It seems to me that, rather than protect the shareholders while we were investigating what happened, the SEC inquiry had the opposite effect. This is unfortunate and needs to be looked into. All this brought me into contact with professions I never knew existed, such as forensic auditors. In addition, the company was spending a lot of money to overcome the burden of some very poor record keeping. This was made even more difficult by the loss of files during the transfer of the audit department from Colorado to New York.

To cap all of this off, the board and I were told in the last quarter of 2002 that we had enough cash to carry us through the next few years. At the same time, the military buildup in Iraq helped keep one airplane flying, and so from the point of view of the board, we appeared to be in good shape. The board requested that we look at the books more deeply, however, as we considered that the cash situation could get very serious in 2003. The board and I requested that the CEO work with our banks and leasing companies to restructure the business financially. At the same time we asked Jeff Erickson, the COO, to concentrate on cost reductions in the company. Unfortunately, communication broke down between the CEO and the board, and this forced the board to bring in a new CEO.

All of this has convinced me that the job of a non-executive chairman is almost impossible unless he has his own staff to check most of the data being given to him. Either that or he must have a CEO who communicates fully. On the other hand, the chairman must remember he is not running the company, only the board. This is a very difficult if not impossible prospect for someone who has previously been a CEO.

During the period from January 2003 to February 2004, Atlas made great strides in restructuring its debt and leases to get into a better position to go into a pre-arranged Chapter 11. This should enable the company to improve as the economy improves. For me, while this episode was both exciting and frustrating, I certainly learned a lot. I would have been far more satisfied, however, if I could have helped them prevent the problems that hurt a lot of people in 2003. Atlas Air also taught me that, if a business goes from a privately held entrepreneurial company to a public company, different kinds of people are needed to run it. This is just like the fact that the people who do the detail design must be different and more disciplined than the preliminary designers.

B/E Aerospace

I joined the board of B/E Aerospace in the middle of 1995. This is a company run by Amin Khoury and his brother Robert, who were in the process of building the company into being a big player in the airline seating business. This was a very small board that included a Swiss entrepreneur engineer, a professor from Harvard, an MBA businessman with his own small business that was doing quite well, and me as the four outside directors. This was an exciting experience, particularly with this small board, and so also was the process of helping the business through the problems involved in integrating some of the new acquisitions they made.

While the Swiss gentleman and I sometimes had heated words over our approaches to the problems we saw, I think we both felt that more order could be put into the production lines and more consolidation could take place. After several missteps, we brought in a Japanese flow expert. B/E Aerospace made many improvements, and now the business has turned around and is ready for the airline rebound whenever it happens.

While the future will be greatly influenced by the sale of new airplanes, 50% of B/E Aerospace's sales are for renovating older airplanes. That part of the business should be able to weather most economic storms providing it keeps overhead low and productivity high. B/E Aerospace has also acquired some component manufacturing companies on the West Coast, which should add some substance to the company, and recently got involved in reconfiguring passenger airplanes into freighters by buying an engineering company in Seattle that has great credibility in doing this work. This whole organization is very forward thinking about what to do and where to go, and now, in addition to seats for passenger airplanes, they produce a lot of aircraft interior components, such as ovens and other things. They also do quite a bit of work on business jets and commuter airplanes, a good business for the future.

The company has a lot of potential but, like most aerospace companies, will have to learn to live with the cyclic nature of airplane orders. Of course by not being completely dependent upon new airplanes, the company should have greater resilience in hard times. During the latest downturn, the Khoury brothers and their team have done a great restructuring job and have reduced the number of facilities considerably. This should put the company in an excellent position when the market recovers.

Textron

In October of 1995, I received a surprise phone call from Barbara Preiskal, who was a GE director. She asked me if I would consider joining the Textron board. This appeared to be a great opportunity for me as it was a prestigious company, and I felt very honored. After being interviewed by the chairman/CEO and several other board members, I joined the board in December 1995. Textron is a very diverse company with business units such as Cessna, Bell Helicopter, auto products, golf carts, and grounds equipment for various golf courses. They are also very big in the fastener business. Joining this board gave me an opportunity to get close to two pieces of an industry which I really liked, the business jet segment with Cessna and the helicopter segment with Bell. Actually this board turned out to be one of the most interesting because of the people on it and the diversity of views as to what was and what was not good for the company. There was a lot of healthy discussion.

Unfortunately, the CEO never felt he had the confidence of the board, although I always got the impression that the board wanted to trust him, and they gave him quite a bit of freedom. This all came to a head when the CEO presented his plan to join with another company to the directors, and it was obvious he had not done his homework, either inside the company or with the investment bankers, as everybody came out against it. When the CEO retired, Lewis Campbell was selected as chairman. A well-balanced board who have a good mixture of experience assist Campbell. Textron is experiencing a complete turnaround under its new chairman and should continue to grow.

During the business downturn of 2002–2003, Lewis Campbell, with the help of his new CFO, started a campaign to transform the company into a premier multinational corporation by building on its many good brand names. They have instituted a series of process improvements to strengthen financial reporting, program management, and manpower development. Most importantly, they have instituted a Six Sigma program that is already producing positive results.

It is very interesting to watch the evolution of organizations when a new chairman comes in. Some people who seemed to be doing a good job under the former chairman began to falter under the new one. Perhaps it can simply be attributed to bad chemistry, but often quite a few heads roll when the new regime takes over. Usually, this is economically painless for those involved, but it must be more than a little frustrating to have worked hard to build a company to a place of prominence only to realize that, because someone does not like the color of your eyes or something, you have to leave. But that is life.

Canadian Marconi

While I was on a business trip to England, I came across an old friend who worked for General Electric Company of the United Kingdom. During our conversation, he asked me if I would be interested in joining the board of Canadian Marconi. GE, U.K., owned a major share of Canadian Marconi; the rest of it was in the public's hands. This company was very interesting in that it was a top-notch electronics company and was associated with the aircraft industry. Another interesting thing about Canadian Marconi was the amount of cash they had. They had not really used it to grow themselves in a period when they probably could have used the leverage of their mother company and grown to be four or five times their size very quickly. I think their failure to seize that opportunity was a real mistake. The British General Electric Company decided they did not want to spend all of the cash, which was up to them, but they also did not grow.

Canadian Marconi's board meetings were held in Montreal, and while these meetings were interesting, I always felt that GE, U.K., made all of the decisions in a meeting prior to the board's meeting with Marconi management. Unfortunately, flying all the way from Florida to Montreal in wintertime for the board meetings really became trying. I resigned from this board in March 1999. Having built some good relationships, I can say I honestly miss working with several of the people there. In fact, I still get occasional letters and telephone calls from them. Canadian Marconi has now gone completely public, and it will be interesting to see how it grows.

Cincinnati Bell

In 1996, Charlie Mechem asked me if I would like to join the board of Cincinnati Bell. This company has always been fascinating to me because it is one of the few original Bells that was not completely attached to AT&T prior to their divestiture, and so it was always run pretty independently. As a result of their independence, Cincinnati Bell had formed a couple of information systems companies—called Matrix and CBIS—within Cincinnati Bell. Matrix did the telephone answering for various technology and telephone companies and had call centers all over the United States, and CBIS was the billing part of the company that processed a tremendous number of telephone bills, not only for Cincinnati Bell but also for many other telephone companies. Their computer programs to do this were really impressive. This was also a very interesting company because people who had been in the business for a long time, but apparently had not been too close to the board, ran it. They brought in Jim Orr from Procter & Gamble to run the CBIS/Matrix part of the company. It soon became obvious that by spinning off both CBIS and Matrix into a new company—Convergys—the stock price of Convergys would be much higher than the stock price of Cincinnati Bell. I was against this idea to start with, but I was convinced in the long run that staying with Cincinnati Bell would stunt the growth of both CBIS and Matrix. The spin-off has worked out well for Convergys but not so well for Cincinnati Bell. Subsequently, Convergys acquired many other companies. When we split the business, I was lucky to be put on the Convergys board. This has turned out well from my point of view, and I hope I have helped contribute to the growth of this company from essentially the first day it was formed until my resignation from the board on my 71st birthday.

Dynatech/Acterna

In the summer of 1998 I got a call from Chuck Pieper, a friend who formerly worked at GE but who was then with Clayton, Dubilier & Rice (CD&R). CD&R was acquiring a company called Dynatech, which had among its various businesses a company called AirShow that does the maps showing where you are when you fly on airplanes. I expressed an interest, and Chuck asked me to join the board. My schedule was already pretty loaded, but he assured me there would be only four or five board meetings a year somewhere in Massachusetts or New York, relatively close to where I live half the year. I accepted. I was there only a couple of months when a problem came up with the chairman of the board, requiring CD&R to put their own man, Ned Lautenbach, in that role. Ned made a big difference in Dynatech. He came from IBM and was a very structured individual. Most importantly, he knew the business world very well and added much wisdom and direction as Dynatech grew.

Dynatech's main business was designing and producing testing equipment for various telephone, cable, and optical systems. With some strategic acquisitions, Ned and the CEO, John Peeler, grew this part of the business to over $1 billion.

Dynatech also made an interesting acquisition of a German company. In the process, they made, in my opinion, a fundamental error. They failed to put a non-compete clause in the agreement. As a result, the former German owner stabbed them in the back by recruiting some of the business' key people to form a competing venture. When you acquire a company—especially a technology company—you must ensure you receive not only the hard assets, but also the brain power, the technical knowledge, and the customer connection of the people who are part of that company. The German acquisition was a big lesson.

This is a very interesting, high-powered board to work on. It can boast of several people who have been chairmen or presidents of various companies.

In 2000, the company name changed to Acterna Corporation. It was really hurt by downturns in the telecommunication industry in 2001, however, and has had its work cut out for it to restructure. One lesson I had previously learned about restructuring was that you must cut quickly to your base requirements, as it is much easier to build up later than suffer through a lack of cash.

They did that, but the telecom industry really suffered from over-expansion combined with rapidly evolving technology. These factors, plus the WorldCom situation, caused a tremendous downturn in the industry that hurt companies that were overleveraged. The problem is that, while a business is good and growing, it needs all of the financial help it can get. Unfortunately, while the people at Acterna reacted really quickly to the downturn in business, they ended up owning just too much debt for the size of the company.

Fairchild-Dornier

During some discussions with UBS, Carl Albert and Jim Robinson of Fairchild-Dornier gave us a presentation on potential acquisition of or investment in that company, which they had bought a couple of years earlier. This was an interesting German company that does work for Airbus as well as builds its own small airplane, the Dornier 328, in both a turboprop and a turbojet version. They had just gotten excited about building a new airplane called the 728, which is a twin-engine, 70-passenger, commuter airplane. On paper, it looked to have great potential; however, when we reviewed the business plan with UBS, and I pointed out the break-even time would be seven to eight years at best and most probably longer, they quickly lost interest. Then Carl Albert went to see Clayton, Dubilier & Rice. Because of my association with CD&R, they asked me to tell them what I thought the positives and negatives of the business were. There is no question that the potential of this airplane was good. Its main competitor is the Embrarer airplane, although, at the time, Bombardier was also looking at a similar airplane. Fairchild-Dornier had obtained a nice order from Lufthansa for 60 airplanes. As I had told UBS, I told CD&R that, if they were willing to stick with this project and if they could make sure they had enough funding up front so that they would not get halfway through and run out of money, this could be profitable for them. They would have to be willing to hang in there for a long time before they broke even, however. Developing a new airplane is really expensive, and people want to believe that nothing will ever go wrong. Unfortunately, bad things do happen. Engineering still involves a lot of art, even with all of these great computer programs that can do so many more things so much faster and more accurately than we ever could in the past. Problems still come up. (As a side note, GE and Fairchild-Dornier agreed to use GE engines on the airplane before I joined the board.)

When Chuck Pieper asked me, after they decided to buy it, if I would like to join the board, I agreed, although I did so reluctantly. I thought it would be a great airplane, but I also felt that it would be too expensive for the market and that they did not have enough funding to complete the development. In addition, board membership meant onerous overseas travel for me. Unfortunately, all of my worries about cash flow were realized. We were in the middle of renegotiating financing when the project announced they would have to delay the critical first flight by four months. This meant we would need even more cash if we were to keep the project going. We could not obtain it—and Fairchild-Dornier had to close. This was very tough on the engineers, but it was even tougher on the investors.

Working with my son

One of the more exciting things I have done since retirement has been working with my son, David, to create a small investment company called Aero Equity. We mainly look at small companies that we think have potential as well as work with UBS and Carlyle on some investments they make in the aerospace field. One area we help in is on the due diligence side where, with the help of TK Engineering, we look at the operations aspects of various businesses to see if they mesh on the financial side. TK Engineering is a company started by Bob Turnbull and Lee Kapor, who were important members of my Aircraft Engines team. They are doing very well in supplying support when required to NASA, GE, and others.

Being on boards

As I look at my roles on the boards I served, the most important contribution I think I made was to get management focused on systems for the development of people, the generation of a much better financial system, and implementation of an orderly management process. I am amazed by how many companies there are, whether large or small, that do not have the management systems they need to grow and prosper. A clear system of formal management processes cuts a lot of unnecessary bureaucracy out of a company. Additionally, such systems allow everyone to know when various elements of the business process are due so that they can all support a consistent business and strategic plan.

It is also very important to have an effective personnel development program. I am concerned when I see people going outside a business to recruit managers who do not understand the culture of the company. Sometimes this has been a valuable strategy, as it was with IBM, but generally people who do not understand the technology or the culture of a company are often more a problem than a solution. One of the things we always did at GE was to try to pick the best people to start with and then grow them to perform bigger jobs. While we've had several good people who came from outside the company, most of the time we were more successful developing GE people. I have tried to suggest this strategy to many of the boards I worked on. Unfortunately, some people reject out-of-hand what they feel to be preaching about the GE way.

Jack Welch had been touted in the popular press as some kind of business god. This often left me appearing to be a prophesizing disciple. While I do not care for the role, frankly, if by being perceived as a disciple I felt I could get these companies operating more efficiently and less bureaucratically, I really enjoyed that. I also enjoyed being on the boards of companies with good technical content so that I could bring some engineering expertise to the development of programs and people—and the technology to keep those programs and people running smoothly.

There is no clear-cut answer to the question of why one person makes a good board member and another does not. I think the thing that I bring to these boards is the fact that I'm willing to ask questions—and not the rhetorical kind. During my leadership career, I was not considered to be one of the best financial people. I was always able to delegate the financial side, and luckily, at GE I always had great financial people working with me. As a result of my exposure to other people's financial expertise, I get asked to be on audit committees, which always surprises me considering my background. I really don't enjoy the financial aspects of audit committees. I find, however, that an audit committee is not all about numbers. It is about asking the right questions, especially with regard to procedures and ethics, and making sure people are *using* the right procedures. As a result of this approach, I've had some great learning experiences, and while I hate to admit it, I really enjoyed my audit committee assignments.

On reflection, my career both at GE and as a director has been very exciting and interesting. I am pleased that I had the opportunity to do many things. I am convinced, however, that the operations side of business is better suited for the young. There is no question in my mind that, while you might not necessarily lose significant vigor or mental capacity as you get older (although obviously some people do), what you do lose is time. A younger man has a lot more time to achieve things than an older man does. People say, "What are you talking about, Brian, time is time is time." My theory of time, particularly in business, is that subjective time is directly proportional to the time you have already experienced. Each successive day, as you grow older, is a smaller percentage of the total number of days in your life. As a result, days go by ever more rapidly and the time within them for doing things seems to dwindle, no matter how scrupulously you keep your calendar. The days that seemed endless containers of doing at age 30 speed by at 70 at more than twice the pace. I sit back now and think of what I used to do in a day at age 40—twice what I can do today, I'm sure.

I regret none of it, though. Business life has been a great experience for me. I have learned a lot—and I am still learning. I can only hope the boards I participated on learned from me as well. I am thankful to all of the people who invited me to join their boards. I grew personally from every one of these experiences, but I hope I contributed more than I gained.

Epilogue
The Propulsion Hall of Fame

When I took charge of GE Aircraft Engines in October 1979, I was very honored to have the job, but I knew that, if it hadn't been for people like Sir Frank Whittle, Dr. Hans von Ohain, and Major Frank Halford in the beginning and many others such as Gerhard Neumann and Jack Parker, as well as people at Rolls-Royce, Bristol, deHavilland, Westinghouse, Curtiss-Wright, and Pratt & Whitney, we would not even have a business. I felt it would be a good idea to honor the people who helped make GE Aircraft Engines what it was and has since grown to be—not only the engineers, but the finance and business people as well as the airplane designers who learned to use our engines. As a result, my staff and I created our Propulsion Hall of Fame, which is formally housed at the GE Aircraft Engines' Evendale facility. The inaugural induction was made in the early 1980s, and the ritual continues to this day with biannual induction ceremonies.

As I mentioned several times in this book, without creative people in nearly all technologies, we would not be producing engines today that are less noisy and more maintainable, with lower fuel consumption and greater reliability. We would not have the current family of engines for the airlines of the world that are carrying more people and bringing the world together. Unfortunately, as we have seen as recently as 9/11/01, our products can be used for things other than peaceful transportation, and many of us in the industry feel very deeply about the misuse of our products. It is also unfortunate that, as good as our products are, they still have accidents, but because of the great engineers who are dedicated to do the detective work to find the causes, we also learn from these accidents to make our products even better. We engineers are a strange breed—sometimes very creative and sometimes very defensive—but none of us likes to feel that our products can be involved in an accident causing people to suffer.

I think that, overall, those involved in our business have done more for the total well-being of the world than all of the sports heroes of tennis, golf, football, and baseball; and of course, each of those sports has its Hall of Fame. In my opinion, it takes not only skill derived through training but also actual creativity and leadership to design our products and put them into use. The same can be said for civil engineers, nuclear engineers, electronic engineers, and ship builders. It is true that there are ways that some of these people are recognized, such as by election to the National Academy of Engineers, but generally the majority of people elected to these honors are academics. I have nothing against academics, but the men and women who design and make actual products seldom get just recognition. As a result, I felt it important that we at GE Aircraft Engines do something to recognize the people who really make our business tick.

As I said many times in this book, it is all about teamwork—not only among engineers, but also throughout the entire organization. In the history of GE Aircraft Engines, there are many people from finance, human resources, manufacturing, sourcing, and product support who deserve to be recognized as well,

With Sir Frank Whittle, left, and Hans von Ohain, right, at GEAE Engineering Recognition Day, February 1985.

and so our Hall of Fame includes all of the professions that helped us achieve the success we have achieved.

I was thrilled to hear Sir Frank Whittle, who had received so many other awards, say how honored he was to be elected to our Propulsion Hall of Fame.

A listing of Propulsion Hall of Fame inductees from 1982 to 2003

1982

Gen. René Ravaud
Sir Frank Whittle

1983

Gerhard Neumann
Jack S. Parker

1984

Bruno W. Bruckmann
C.W. LaPierre
Clarence E. Danforth
Donald F. Warner
Dr. Sanford A. Moss
Edward Woll
Eugene E. Stoeckly
Fred O. MacFee Jr.
Joseph S. Alford
Kenneth N. Bush
Morrough P. O'Brien
Perry T. Egbert
Robert B. Ingraham
Samuel J. Levine
Virgil L. Weaver
William L. Badger

1985

Bruce O. Buckland
Dale D. Streid
David Cochran
Donald C. Berkey
G. William Lawson
Glenn B. Warren

J. Walter Herlihy
John F. Klapproth
Louis P. Jahnke
M. Robert Rowe
Neil Burgess
R. Walter Brisken
Theodore J. Rogers

1986

Everett J. Kelley
Leander J. Fischer
Lt. Gen. Laurence C. Craigie
Nicholas J. Constantine
Peter G. Kappus
Roy E. Pryor
Walter S. Bertaux

1987

John W. Jacobson
Kenneth O. Johnson
Martin C. Hemsworth
Nicholas F. Frischhertz
Richard W. Hevener Jr.
Walter F. Cronin

1989

Arthur P. Adamson
Bert E. Sells
Fred I. Brown Jr.
Frederick W. Garry
James N. Krebs
Ned A. Hope
O.R. (Bud) Bonner
Ralph E. Patsfall

1991

George H. Ward
Jackson R. McGowan
John W. Blanton
Richard B. Smith
Robert C. Hawkins
Robert D. Desrochers

1993

Frank R. Homan
Harry LeVine Jr.
Irving W. Victor
James E. Worsham
Melvin Bobo
Robert E. Neitzel
Stephen J. Chamberlin
Walter E. Van Duyne
William J. Crawford III

1995

Anthony J. Nerad
Brian H. Rowe
Calvin H. Conliffe
Donald F. Keck
Donald W. Bahr
Edward E. Hood Jr.
Frank E. Pickering
Fredric F. Ehrich
Lee Kapor
Leroy H. Smith Jr.
Lester H. King
Michael J. Stallone
Robert C. Turnbull
Robert J. Gerardi
Thomas F. Donohue
W. George Krall

1997

Bernard J. Anderson
Dean J. Lennard
Henry J. Brands
Larry A. Scott
Louis V. Tomasetti
P. Arthur Adinolfi
Walter B. Houchens

1999

Barry Weinstein
Brian Brimelow
Edward C. Bavaria
John T. Moehring
Peter A. Chipouras
Pierre Alesi
Robert J. Smuland
Robert L. Sprague

2001

Bruce Gordon
James E. Sidenstick
James Heyser
Robert B. Kelly
William E. Schoenborn

2003

Louis A. Bevilacqua
Vincent M. Cardinale
Samuel H. Davison
Paul Joseph Hess
Vincent LaChapelle
Georges Sangis
Theodore T. Thomas Jr.
James C. Williams

INDEX

A-10 Thunderbolt, 32, 35
A-Bar-A Ranch, 171
A300. see Airbus Industrie, A300
A300-600. see Airbus Industrie, A300-600
A310. see Airbus Industrie, A310
A320. see Airbus Industrie, A320
A330. see Airbus Industrie, A330
A340. see Airbus Industrie, A340
A380. see Airbus Industrie, A380
Abraham, Rheinhardt, 172
Acterna, 189–190
Adams, Al, 172
Adamson, Arthur P., 22, 26, 32, 51, 79, 195
Adinolfi, P. Arthur, 32, 196
Advanced Engineering, 33, 39
Advanced Projects, 26–27
Aegis, 77
Aero Equity, 191
Aerostructures Hamble Ltd., 183
Air Force One, 56
Air France, 52, 57, 74, 145
Air Holdings Ltd., 48
Air India, 57
Air New Zealand (ANZ), 57, 151
Airbus Industrie, 46, 54–55, 57–58, 69, 71, 74–75,
    127–128, 134, 146, 152, 154, 172–173, 190
    A300, 54–56, 69–70, 152
    A300-600, 58
    A310, 57
    A320, 74, 152
    A330, 58, 75, 123, 134
    A340, 75–76, 123, 127, 135
    A380, 146
    Airbus Jet 1, 71
Aircraft Engines. see General Electric (GE), Aircraft
    Engines
Aircraft, crew, maintenance, and insurance (ACMI)
    model, 184
AirShow, 189
Albert, Carl, 190
Alesi, Pierre, 76, 196
Alford, Joseph S., 138, 195
Alien Office, 20–21
Alitalia, 52, 171
All Nippon Airways (ANA), 57–58, 130, 145,
    156–158
Allied Signal, 153
American Airlines, 40–41, 46–47, 49, 132, 145
Anderson, Bernard J., 196
Anderson, Roy, 56
Andrews, Wayne, 156
Apprenticeship, 6–9
ARJ21, 146
Armed Services Committee, 66–67
Armstrong Siddeley, 15

Armstrong, Neil, 168
Army Air Corps, 59
Arthur Andersen Audit Company, 185
Asakura-san, 157
AT&T, 188
ATLAS, 50, 52, 57
Atlas Air, 184–186
Atlee, Clement, 149–150
Australia, 20, 78, 168
Avon, 41
Axial-flow turbojet, 14–15
Axicentrifugal flow engine, 31
Ayling, Bob, 161

B-1, 69, 71, 150
B-17, 21
B-2, 67
B-36, 59
B-45, 59
B-47, 59
B-58, 59
B-70 Valkyrie, 43, 45
B/E Aerospace, 186–187
Badger, William L., 195
Bahr, Donald W., 196
Barrett, James, 130, 160
Battle of Britain, 1–5
Bavaria, Edward C., 58, 83, 122, 130, 145, 157, 168,
    196
Begin, Menachem, 152
Bell Helicopter, 187
    Bell 714, 34
Benchmarking, 108–109
Benichou, Jacques, 91
Benzakein, Mike, 76, 143, 145
Berkey, Donald C., 25, 45, 195
Bertaux, Walter S., 195
Bevilacqua, Louis A., 196
Bilien, Jean, 76
Bird tests, 141
Bisignani, Giovanni, 171–172
Black Belt, 111
Blair, Tony, 161
Blanton, John W., 196
Boating, 182
Bobo, Melvin, 76, 138, 196
Boca Raton, Florida, 93, 95
Boehm, Ted, 117, 122
Boeing, v, vi, 17, 28, 41–42, 45–47, 50, 54–58, 64,
    71–73, 90, 95, 127–130, 132, 135, 139–140,
    142–145, 152–154, 157–158, 161–163, 172,
    184–185
    707, 41, 45, 56, 71–72, 161
    727, 46–47, 80, 158, 162
    737, 73–74, 90, 92, 129–130

737-200, 73
737-300, 152
747, v, 45–47, 50, 55–58, 130–131, 140, 151, 157, 163, 184–185
747-300, 58
747-400, 58, 130, 157, 184
747SR, 156
767, 57–58, 71, 156
777, 64, 123, 128–132, 135, 139–140, 143–145, 157, 163
777-200LR, 145
777-300ER, 145
B-47 Stratojet, 17
Stratocruiser, 28
Bonner, O.R. (Bud), 78, 195
Borger, John, 27–28
Boston, Massachusetts, 23–24
Boxford, 24, 34
Boyne, Bob, 53
Brands, Henry J., 196
Brigham, Paul, 100
Brimelow, Brian, 122, 196
Brisken, R. Walter, 195
Brisken, Tom, 76
Bristol, 193
  Blenheim, 4
  Olympus, 24
  Siddeley, 15–16
British Aerospace, 183
British Airways (BA), 57, 130–133, 135, 143, 160
British Caladonian, 172
British Overseas Airway Corporation (BOAC), 40
British Science Museum, 7–8
Brizendine, John, 49
Brodie, John, 12
Brooks, Arnie, 29
Brown, Fred I., Jr., 195
Brown, Wayman, 173
Bruckmann, Bruno W., 195
Buccaneer, 15
Buckland, Bruce O., 195
Buenger, Clem, 182
Burgess, Jack, 40
Burgess, Neil, 151, 195
Burk, Dick, 101
Burlingame, John, 88
Bush, George H.W., 158–160
Bush, Kenneth N., 195

C-5 Galaxy. see Lockheed, C-5 Galaxy
C-5A. see Lockheed, C-5A
Calhoun, Dave, 145
Callaghan, 77
CAMMACORP, 71
Campbell, Lewis, 187–188
Canadair, 35–36
  Challenger, 32, 35–36, 64
  Regional Jet, 32, 64
Canadian Marconi, 188

Canberra bomber, 17
Cape Cod, 131, 156, 181
Cardinale, Vincent M., 196
Carlyle, 191
CASA 212, 154
Caterpillar, 102
Cathay Pacific, 41, 57
CBIS, 189
Center of Excellence, 112–113, 123–124
Cessna, 187
CF34, 36, 64, 92, 110, 123
CF34-10, 146
CF6, vii, viii, 36, 40, 46–58, 64, 67, 69–70, 75–76, 79–81, 83, 101, 123, 127–129, 132, 134–135, 145, 151–152, 157, 162
  Nacelle, 48–49, 54
CF6-50, 50, 52, 55–56
CF6-6, 47–51, 57, 81
CF6-80, 57–58
CF6-80A, 58
CF6-80C, 58, 76, 132
CF6-80E, 123
CF6-80E1, 134–135
CF700, 27–29, 42, 64, 100
CFM International, 71–72, 74, 91
CFM56, 64, 66–67, 70–76, 90, 121
CFM56-5, 74–75, 92, 123
CFM56-5C, 75, 127
Chadwell, Chuck, 85, 101, 122, 145
Chamberlin, Stephen J., 196
Chelsea, 165
Chennault, Anna, 82
Chennault's Flying Tigers, 32
Chevrolet, 59
Chicago, Illinois, 20, 130
China, 32, 78, 146
Chipmunk, 8
Chipouras, Peter A., 196
Chowdry, Linda, 185
Chowdry, Michael, 184–185
Chula Vista agreement, 49
Churchill, Winston, 149–150
Cincinnati Airport, 35
Cincinnati Bell, 188–189
Cincinnati Gas and Electric, 77
Cincinnati, Ohio, 19–20, 23, 33–34, 57, 59, 77, 81–82, 98–99, 158, 160, 167, 181–182
CJ610, 64
CJ805, 41, 52, 64, 157
CJ805-23, 27–28, 42, 52
CJ805-3, 59
Clapper, Bill, 122
Clayton, Dubilier & Rice (CD&R), 189–190
CN235, 153–154
Cochran, David, 24, 195
Cold War, 120, 134
Collier Trophy, 80
Colodny, Ed, 73
Comair, 35

Index

Comet, 10, 12–14, 16, 40
Commerical Engine Operation, 145
Commerical Engines Operation, 176
Compartmentalization, 108
Composite blade, 137–138
Conaty, Bill, 85, 117, 125
Concorde, 69, 91
Conliffe, Calvin H., 196
Conquistadores del Cielo, 151, 171–174, 176, 184
Constantine, Nicholas J., 195
Construcciones Aeronáuticas SAC (CASA), 153
Continuous Improvement Program, 92, 94, 98, 104–105, 109–113, 121, 123
Convair, 41
    Convair 880, 41, 59, 64, 157
    Convair 990, 27, 41, 51, 64
Convergys, 178
Cooper, Tom, 122
Cordiner, Ralph, 27
Craigie, Laurence C., 195
Crawford, William J., III, 196
Cronin, Walter F., 195
Crotonville, 87, 93–95
CT58, 64
CT7, 34–35
Curtiss-Wright Aeronautical, 40, 193

Daimler Aerospace (DASA), 133–134
Danforth, Clarence, 195
Dassault, 28–29
    Falcon, 28
    Mystere 20, 28
David, George, 161
Davis, Morrie, 151
Davison, Samuel H., 196
deHavilland Engine Company, 6–17, 19, 21, 59, 98, 193
    Ghost, 10, 12–14, 16, 37, 40, 98
    Gipsy Major, 8
    Gipsy Minor, 8
    Gipsy Queen, 8
    Goblin, 10, 12, 14, 37
    Spectre, 16–17
    Sprite, 16
    Vampire, 10, 12
deHavilland, Sir Geoffrey, 13
Delta Airlines, 41, 49, 57, 71, 145, 154
Deming, W. Edwards, 104, 108–110, 121, 157
Department of Defense, 42
Desrochers, Robert D., 34, 83–84, 196
Destroyer, 77–78, 135
Diedrich, Gunther, 19
Dinsmore, John, 181
Dodge, Wally, 19, 57
Dolfi, Sam, 85
Donohue, Thomas F., 138, 196
Doorly, Eric, 34
Dornier 328, 190
Doster, John, 183
Dotan affair, 115–116

Dotan, Rami, 115–116, 152
Douglas, 42, 46, 154
    DC-6, 98
    DC-8, 41, 45, 71–72, 74
    DC-10, 40, 47–49, 52, 54, 129, 156, 158, 162
    DC-10-10, 50, 54, 81
    DC-10-20, 52
    DC-10-30, 50, 52, 55–56, 81, 151
Downsizing, 119–120, 124
Duncan, Sir William, 132–133
Dunkler, Herr, 133
Durham University, 13–15, 59, 166
Durham, North Carolina, 124
Dynatech, 189–190

E-4, 56
E^3 (Energy Efficient Engine), 135
Eastern Airlines, 46, 48–49
Edgeware, 1, 3, 6, 9
Edwards Air Force Base, 22
Edwards, Jack, 154
Egbert, Perry T., 195
Ehrich, Fredric F., 196
Electrosteam drilling, 99, 104
Embraer 190, 146
Embraer 195, 146
Encampment, Wyoming, 171
Energy Efficient Engine. see E^3 (Energy Efficient Engine)
Engine tests, 141
Engineering Department, 168
Engineering Division, 111, 177
England, 1–2, 9, 20, 131, 165–166, 178, 183
Environmentalists, 141, 161–163
Erickson, Jeff, 186
Ernst & Young, 185
EVA Airways, 145
Evendale, Ohio, 19, 23, 33, 37–39, 81–82, 101, 123–124, 151, 158, 177
Extended twin operation (ETOP), 139–140, 142

F-4 Phantom, 47
F-5 Freedom Fighter, 31
F-14 Tomcat, 39, 66, 151
F-14G Super Tomcat, 67
F-15 Eagle, 39, 66, 151
F-16, 66–67, 115, 150–152
F-86 Sabrejet, 59
F/A-18 Hornet, 37, 60, 66
F100, 37, 39, 67, 150–151
F101, 64, 66, 69, 71
F101-X, 66
F101DFE (Derivative Fighter Engine), 66
F110, 67, 83, 115, 124
F118, 67
F136, 146
F404 engine, 37–38, 60, 64, 66, 79–80
F414, 37
FAA, 34, 63–64, 71, 139, 141

Fairchild-Dornier, 190–191
Fairfield, 36–37, 83, 87–88, 94, 101, 154
Falcon, 28
Far East, 82
Father, 2, 4–5
FedEx, 36
Fiat, 74
Fiat-Avio, 133
Fifth Third bank, 182
Film cooling, 43
Fischer, Leander J., 195
Fleming, Leslie, 181
Flight Propulsion Development, 24
Florida, 78, 93, 188
Flying Tiger Line, The, 71
Fokker, 73
Football (American), 168
Forsyth, Charlie, 172
France, 3, 54, 69–72, 90–91, 139, 168
Fresco, Paulo, 89
Frey, Mr., 171
Frischhertz, Nicholas F., 195

Garrett, 34
Garry, Frederick W., 30, 33, 195
Garuda, 41, 154
Gault, Stan, 88
GE1, 23–25, 39, 44
GE1/6, 39, 44
GE12, 31–32, 34
GE90, v, ix, 11, 64, 76, 80, 83, 102, 121, 123–124,
    127–129, 131, 133–147, 157, 160
GE90-115B, 145–146, 157
General Dynamics, 67
General Electric (GE), iv–xi, 8–9, 11, 17, 19–30,
    32–37, 39–43, 45–52, 54–57, 59–61, 63–64,
    66–67, 69–79, 81, 83–84, 87–96, 98, 100–101,
    103–104, 109, 111, 113, 115–117, 119–121,
    123–125, 130–133, 140, 142–143, 145,
    150–152, 154–161, 167–169, 172–173,
    176–178, 181–185, 187–193
    Aircraft Engines, 19, 32, 34, 37, 40, 72, 77–79, 83,
        85, 87–90, 93, 101, 111, 113, 116–117,
        120–124, 131, 143, 145, 159, 173, 176–178,
        182–183, 191, 193
    Field Engineering, 157
    Military Engine Operations, 37, 115
    Capital Aircraft Services (GECAS), 140
    Corporate Executive Council, 93
    Engine Services, 130
    Plastics, 88
General Motors
    Allison Division, 49
Gerardi, Robert J., 122, 196
Germany, 3–4, 20–21, 32, 54, 133, 153–154, 158,
    189–190
Ghost. see deHavilland Engine Company, Ghost
Gipsy Major, 8
Gipsy Minor, 8

Gipsy Queen, 8
Gloucester Gladiator, 3
Goblin. see deHavilland Engine Company, Goblin
Goldsmith, Bob, 82–83, 176
Golf, 157, 168, 182, 187, 193
Gordon, Bruce, 40, 127, 143, 196
GP7000, 146
Gradison, Bill, 159–160
Gray, Harry, 161
Great Engine War, 66–68, 150–152
Greenhills, 24
Gregor, George, 156
Grimes, Mike, 183
Grumman, 67
Gulf War, 32, 35, 120, 134, 158
Gulfstream, 36
Gunboat, 77
Gyron, 14–15
Gyron Junior, 15, 17

Habibie, Ainun, 153
Habibie, Rudi, 153–156
Hagrup, Knut, 50
Haig, Alexander, 161
Halford, Major Frank, 12, 193
Hampton, Tom, 67
Harrington, Lou, 154–155
Harris, Hollis, 168
Hatfield, 16
Hauser, Ambrose, 135, 142
Hawker Hurricane, 1, 3, 175
Hawker-Siddeley Harrier, 24–25
Hawkins, Robert C., 39, 79, 168, 196
Heil, Russ, 154–155
Heinkel jet engine, 19
Hemsworth, Martin C., 26, 45, 135, 138, 195
Hendon, 3
Hendon Technical School, 6
Herlihy, J. Walter, 195
Hernandez, Henry, 181
Hess, Paul Joseph, 196
Hevener, Richard W., Jr., 195
Heyser, James, 196
High-bypass engine, 42–46, 128, 137, 161
Hineman, Ben, 117
Homan, Frank R., 76, 196
Honeywell, 34
Hood, Edward E., Jr., 40, 42, 46, 49, 51, 55, 73–74,
    81, 88, 171, 196
Hooksett, New Hampshire, 123
Hope, Ned A., 195
Horan, Jack, 30
Houchens, Walter B., 196
Hubschman, Henry, 117
Hughes, 172
Hughes, Howard, 41
Hurtt, K., 173
Hydrofoil, 77

Index

I-A, 59
Iberia, 52
IBM, 86, 189, 191
Immelt, Jeff, 145
Inaba, Dr., 157
Indonesia, 153–156, 168
Industri Pesawat Terbang Nusantara (IPTN), 153
Inertia welding, 102–103
Ingraham, Robert B., 195
International Aero Engines (IAE), 74–75, 152
   V2500, 74
International Group, 81
Ishikawajima Heavy Industries (IHI), 139, 142–143, 157
Israel, 115–116, 152, 160
Israeli Air Force, 115

J33, 59
J35, 59
J47, 17, 49
J73, 59
J79, 17, 25, 40–42, 47, 64, 66, 77
J85, 21, 24, 26–27, 31, 38, 64, 100–101
Jacobson, John W., 195
Jahnke, Lou, 102
Jahnke, Louis P., 195
Japan, 58, 78, 108, 156–158, 168, 187
Japan Airlines (JAL), 39, 41, 57–58, 130, 145, 156–158
Japanese Aero Engines, 74
Jet engine design, 10–11
Johannesburg, South Africa, 16
John, Elton, 166
Johnson, Kenneth O., 80, 195
Joint Strike Fighter, 146
Jones, Reginald, 81, 88–89, 178
JT3, 41, 72
JT3D, 42, 73
JT8, 71, 80
JT8D, 70, 73–74
JT9D, 46, 52, 55
JT9D-20, 52

Kapor, Lee, 75, 122, 138, 191, 196
Kappus, Peter G., 21–23, 195
Kaye, Danny, 81
KC-135, 66, 72–74
Keck, Donald F., 196
Kelleher, Herb, 73
Kelley, Everett J., 195
Kelly, Robert B., 38, 196
Kennedy, Ted, 160
Kenton Cricket Club, 166
Key Largo, 181
Khoury, Amin, 186–187
Khoury, Robert, 186–187
King David Hotel, 152
King of Morocco, 158
King, Lester H., 196
King, Lord John, 130–132, 143, 160–161, 183

Kings College, Durham University, 13–15
Kirk, Bob, 36
Klapproth, John F., 195
KLM, 50, 57
Knutsen, Ken, 154–155
Koff, Ben, 67
Kolk, Frank, 47
Korean War, 59
Krall, W. George, 53, 57, 83, 101, 122, 124, 196
Krebs, James N., 40, 46, 66–67, 83, 195
KSSU, 50–52, 57
Kutney, John, 28–29

Labour Party, 149
LaChapelle, Vincent, 196
LaGuardia Airport, 47
Lambeth, England, 2
LaPierre, C.W., 195
LaPlante, Cliff, 66
Larudier, Bernard, 28
Laser drilling, 99
Lautenbach, Ned, 189
Lawson, G. William, 195
Leadership Course, 176
Lehman, John, 63
Lenher, Frank, 76
Lennard, Dean J., 196
Lester, Don, 85, 120
Letts, Ray, 82–83
Levesden, 15
LeVine, Harry, Jr., 66–67, 150, 196
Levine, Samuel J., 30, 77–78, 195
Lewis, Dave, 50, 156
Lewis, Linwood, 154–155
Lift fans, 21–23
Ling-Temco, 98
Little, Dennis, 79, 122, 143
Litton, 78
LM100, 77
LM1500, 77
LM1600, 79
LM2500, 77–78
LM5000, 79
Lockhardt, Mike, 122
Lockheed, 17, 41–42, 45–50, 54–55, 59, 62, 69, 156
   C-5 Galaxy, 45, 49, 56, 60, 62, 77
   C-5A, 62
   F-104 Starfighter, 17, 41, 59
   L-1011, 40, 47–48, 50, 54–55, 69, 156
Lockland, 20, 59
London, England, 2–4, 19, 131, 143, 149
Lufthansa, 52, 56–57, 74, 172
Luken, Tom, 160
Lycoming, 35–36
Lynn Production Engineering, 30–32
Lynn River Works, 23
Lynn, Massachusetts, 23–26, 34, 37–38, 59, 82–83, 99, 123–124, 135, 143, 160, 167, 177

M56, 69–70
MacFee, Fred O., Jr., 23, 33, 57, 72, 82–83, 120, 177, 195
Madisonville, Kentucky, 123
Major, John, 160–161
Malroux, Jean-Claude, 72, 76
Manchester United, 166
Manufacturing, 97–105, 193
Marine and Industrial Engine Department (M&I), 77–80
Marshall, Colin, 130–132, 143, 160–161
Master Black Belt, 111
Materials Lab, 102
Matrix, 189
McCord, Sandy, 58, 156–157
McDonnell Aircraft Company, 47
McDonnell Douglas, 39, 47–50, 52, 54–55, 59, 81, 83, 127, 156, 158
    F-4 Phantom, 59
    MD-10-30, 58
    MD-11, 58
    MD-80, 39, 80
McDonnell Grumman F/A-18 Hornet, 37
McGowan, Jackson R., 50, 71–72, 81, 156, 196
McNamara, Robert, 62
McNerney, Jim, 145
Mechem, Charlie, 188
Medros, Ralph, 82
Metzenbaum, Howard, 160
Middlesex, 6
Miles, Bob, 33, 77
Mitsubishi Zero, 32
Mitsui, 156
Mobile, Alabama, 154
Moehring, John T., 196
Mohave, 36
Mojave, 140
Montgomery, John, 40
Montreal, 188
Morocco, 158
Moscow Dynamo, 165
Moss, Sanford A., 59, 195
Motoren-und-Turbinen-Union (MTU), 54–55, 74, 133–134
Motorola, 113
Moxon, Bill, 167
Mueller, Dave, 35
Mulally, Alan, 143
Muldoon, Robert David, 151, 158
Murphy, Gene, 83, 122, 143, 145

N250, 154
NASA, 22, 80, 135, 191
National Academy of Engineers, 193
National Security Council, 153
Navy War College, 78
NBC, 168
Neitzel, Robert E., 196
Nelson, Jim, 104
Nerad, Anthony J., 196
Neumann, Gerhard, 23, 29, 32–33, 39–40, 42, 46, 51, 55–56, 71–72, 77–78, 81–83, 86–88, 91, 120, 176–177, 193, 195
    Neumann's axioms, 86–88
Neutron Jack, 89. see also Welch, Jack
Neville Chamberlain, 3
New York, 186, 189
New Zealand, 151–152, 158, 168
Newcastle, 15, 166
Nietzel, Bob, 40
9/11. see September 11, 2001
Nine Initiatives, The, 109
Nixon, Richard M., 71
Noise, engine, 162–163
North American Aviation, 22
    F-86 Sabre Jet fighter, 17
Northrop flying wing, 59
Northwest Airlines, 52, 57
Norton motorcycle, 12, 16
Norton, Mr., 16

O'Brien, Morrough P., 195
O'Neill, Thomas P. "Tip," 160
Oliver, George, 124
Orr, Jim, 189
Ossimi, Dr., 157

P-80 Lightning, 59
Pakistani International Airlines, 145
Pan American Airlines, 27–28, 45–46, 184
Paris Air Show, 50, 133
Park, Jim, 84
Parker, Jack S., 40, 50–51, 56, 71, 83, 91, 193, 195
Patsfall, Ralph E., 195
Peebles, Ohio, 21, 80, 138–139, 141, 151
Peeler, John, 189
Pétanque, 173
Pickering, Frank E., 37, 83, 122, 196
Pieper, Chuck, 189–190
Pierson, Jean, 75, 172–173
Pinch rolling, 100
Pirtle, George, 142
Pirtle, John, 81, 142
Polar, 185
Politics, 149–163
Pompidou, Georges, 71
Power Systems, 78–79, 83, 88
Pratt & Whitney, 28, 32, 37, 39–42, 46–47, 52–53, 55–58, 60, 66–67, 69–75, 77, 79–80, 90, 95, 104, 120, 129–130, 133–134, 140, 143, 146, 150–152, 157–158, 161, 193
    F100, 152
    F120, 152
    PW4000, 58
Preiskal, Barbara, 187
Prince Phillip, 178
Production Engine Department, 40
Propulsion Hall of Fame, 91, 193–196

Index

Proxmire, Sen. William, 60
Pryor, Roy E., 195

Quiet Engine Program, 136

Ravaud, Rene, 69–70, 72, 90–91, 195
Raymond, Mr., 15
Reagan, Ronald, 178
Reece, Sir Gordon, 160
Reiner, Gary, 113
Rene' 95, 102
Republic Aviation, 22
Republic P-84, 59
Revaud, Rene, 74, 76
Richardson, John, 172–173
Robinson, Frank, 122
Robinson, Jim, 190
Rodenbaugh, Bill, 40
Rogers, Theodore J., 195
Rohr Industries, 48, 54, 154
Rolls-Royce, ix, 40–41, 46–50, 54–58, 69, 74, 76, 104, 129–133, 145–146, 151–152, 160–161, 176, 193
    RB.207, 55
    RB.211, 69
    RB.211-22, 47–48, 55
    RB.211-524, 57
    RB.211-535, 132
    RB.211-Trent, 58, 145, 160
Rowe, Bill, 182
Rowe, David, 81, 140, 169, 181, 191
Rowe, Jill, 19–20, 24, 33–34, 81, 115, 143–144, 167, 177–179, 181, 183
Rowe, Linda, 81, 181
Rowe, M. Robert, 195
Rowe, Penny, 181
Royal Air Morocco, 158
Rutland, Vermont, 100, 123
Ryan Aeronautical, 22
    XV-5A, 21–24

S-3, 35
S-3A, 32, 35
S-duct, 47
Saab, 34
Saab 340, 35
Sabena, 52
San Marcos, Texas, 137
Sangris, Georges, 196
Sapphire, 15
SAS, 41, 51, 505
Saudi Air Force, 183
Saunders Roe, 16
    SR53, 16
Scandinavian Airlines, 50
Schaefer, George, 182
Schenectady, 79
Schoenborn, William E., 196
School of Engineering of Kings College, Durham University, 13–15

Scott, Larry A., 196
Scowcroft, Brent, 153
Seattle, Washington, 57, 187
Secret Service, 158–159
Securities and Exchange Commission (SEC), 186
Sells, Bert E., 195
September 11, 2001, 95, 185, 193
Session C, 94–95
Session I, 94–95
Session II, 95
707. see Boeing, 707
727. see Boeing, 727
737. see Boeing, 737
747. see Boeing, 747
767. see Boeing, 767
777. see Boeing, 777
Shibuya, Ken, 156
Sidenstick, James E., 196
Sikorsky Black Hawk, 32
Singapore Airlines, 57
Six Sigma, 94, 104, 111, 113, 125, 143, 188
Small Aircraft Engine Department, 30
Small Aircraft Engines, 23
Smith, Leroy H., Jr., 196
Smith, Richard B., 52, 72, 76, 81, 196
Smith, Roy, 138
Smithsonian Institution, 7–8
Smuland, Robert J., 79, 196
Snecma, 54–55, 66, 69–71, 73–74, 76, 83, 90, 133, 139, 142–143
Snitzer, Erwin, 20
Soccer, 6, 165–166, 168–169
Sollier, Jean, 76
South Korea, 78, 151
Southwest Airlines, 73
Soviet Union, 134
Spain, 54, 78, 153
Spantax, 41
Sparks, Russ, 143, 145
Sports, 5–6, 165–169
Sprague, Robert L., 196
St. Andrews, 168
St. Croix, 181
Stag Lane, 6, 9
Stallone, Michael J., 196
Steam Turbine Division, 77
Steckley, Gene, 77
Steindler, Herb, 115–116
Stewart & Stevenson, 79, 183
Stewart, Jim, II, 183
Stiber, Bob, 122
Stirgwolt, Ted, 22, 48
Stoeckly, Eugene E., 138, 195
Stonecipher, Harry, 53, 57, 73, 79, 81, 83, 172
Strategic Air Command, 72
Streid, Dale D., 195
Strother, Kansas, 124
Suharto, General, 153, 155
Super-Fan engine, 75

Supercharger, 10, 28
Supercharger Engineering Department, 59
Sutter, Joe, 57
Sweden, 10, 34, 37, 50, 80
Swifton Village, 20
Swimming, 166–167, 182
SwissAir, 41, 50–51
Szecskay, Frank, 157

T39, 40
T58, 24, 30, 62, 64, 77, 100
T64, 30, 32, 34, 38, 100
T700, 31, 34, 38
Tanaka, 156
Tennis, 157, 167–168, 172–173, 182, 193
Textron, 187–188
TF30, 39
TF34, 32, 34–36, 64
TF39, 25, 32, 42–46, 49, 56, 60, 64
Thai Airlines, 41
Thailand, 168
Thatcher, Margaret, 151, 157–158, 161
Thomas, Theodore T., Jr., 196
Thompson Lab, 99
Thompson, Adam, 172
Thornton, Dean, 153
Tiger Moth, 8
Time to Think, 93–94
TK Engineering, 191
Tomasetti, Louis V., 57, 196
Tombs, Francis, 132
Torbeck, Ted, 124
Total Quality, 110, 121
Total Quality Advisors, 109–111
Trippe, Juan, 27
Tucker, Jim, 138
Tulsa, Oklahoma, 40
Turbine Airfoils Center of Excellence, 112
Turbo-supercharger, 21, 59
Turbochargers, 10
Turbofan, 36–37, 42, 46
Turnbull, Robert C., 76, 83, 122, 168, 191, 196
TWA, 41, 46, 48–49, 171

U.S. Air Force, 39, 42, 63, 66–67, 71–72, 77, 115–116, 150–151, 183
    Procurement Office, 67
U.S. Army, 21
U.S. Marine Corps, 25
U.S. Navy, 25, 35, 37, 39, 58, 60, 63, 66–67, 77–79
UBS, 190–191
Unducted Fan engine (UDF), 76, 79–80, 136
United, 71
United Airlines, 46, 49, 86–87, 129–130, 132, 145
United Kingdom, 5, 48–50, 54, 59, 131–132, 149–150, 155, 160–161, 166, 183, 188
United States, 17, 19, 32, 46, 50, 54, 56, 59–61, 71, 81, 90, 115–116, 129, 132, 149–150, 152, 154–156, 158, 166, 168, 178, 184

United Technologies Corporation (UTC), 133–134, 161
US Airways, 73
UTA, 50

V-1, 5
V-2, 5
Vampire. see deHavilland Engine Company, Vampire
Van Duyne, Walter E., 196
Vanderslice, Tom, 88
VanDuyne, Walt, 53
Vareschi, Bill, 122, 131
Varig, 41
Venom, 10
Vertical takeoff and landing (VTOL), 21, 25
Victor, Irving W., 196
Vietnam, 31, 116
Villaroche, France, 72, 139
Viper, 27
Volvo, 37
von Braun, Werner, 20
von Ohain, Dr. Hans, 193–194

Wales, 130–131, 160–161
Ward, George H., 67, 196
Warner, Donald F., 195
Warren, Glenn B., 195
Warthog, 35. see A-10 Thunderbolt
Water polo, 166
Watford Juniors, 166
Weaver, Virgil L., 195
Weinstein, Barry, 196
Welch, Jack, 73–74, 85, 88–91, 94, 96, 109–111, 113, 122, 143, 145, 154, 178, 191
Welsch, Ron, 48, 135, 140
Welsh Development Authority, 160
West Coast, 187
West German Air Force, 16
Westinghouse, 19, 193
Whitbeck, Cliff, 53
White House, 152–153
Whitmore, Phil, 100
Whittle, Sir Frank, 12, 59, 91, 193–195
Wiater, Richard, 183
Wilkens, Mr., 27
Williams, James C., 122, 196
Wilmington, North Carolina, 123
Wilson, Dr. Kerr, 11
Wilson, T., 56, 71
Wiseman, 152
Woll, Edward, 23, 26–27, 37, 66, 76, 82–83, 195
Woods Hole Institute, 155–156
Working Together program, 49, 135
Workout, 109–111, 123
World War I, 59
World War II, 3, 21, 77, 149, 165
WorldCom, 190
Worsham, James E., 33, 39, 45, 83, 176, 196
Wright Aeronautical, 53, 59

Index

Wright Brothers, 7–8
Wright Flyer, 7–8
Wright, Bob, 168
Wright-Patterson Air Force Base, 67

Young, Ambrose, 158

Ziegler, Henri General, 55
Zipkin, Morrie, 66–67, 152

Supporting Materials

A complete listing of AIAA publications is available at http://www.aiaa.org.